Identity Crisis

Also by Jefferson Bass

Identity Crisis

The Murder, the Mystery, and the Missing DNA

JEFFERSON BASS

WITNESS
IMPULSE

An Imprint of HarperCollinsPublishers

EPub Edition MAY 2015 ISBN: 9780062419842
Print Edition ISBN: 9780062419897

10 9 8 7

Valley Authority—were an intimidating reminder of the reason for most of the roads . . . and perhaps many of the gravestones.

On this morning in August 2005, I was threading my way over Graves Gap and down the north side of Redoak Mountain to exhume a grave in a family cemetery in a remote, rugged corner of Anderson County. Three months earlier I'd gotten a call from a young woman named Michelle Atkins. A recent UT graduate, Michelle had heard me guest-lecture in her chemistry class, where I talked about forensic anthropology, my postmortem-research facility called the Body Farm, and a few of the cases I had worked during nearly five decades as a bone detective. When she phoned, Michelle—an avid fan of the forensic drama *CSI*—asked if I might be able to help answer a question that had been troubling her family for more than twenty-five years— ever since the disappearance and apparent murder of her grandmother: Was the skeleton that had been found and identified as her grandmother's, and later buried in the family cemetery, really who the authorities said it was? Or might the investigators have made a mistake in their eagerness to close the case and to put an end to the family's repeated inquiries?

It sounded fascinating. "I'll be happy to help if I can," I told Michelle. She gave me a bit more background, and a few days later one of her aunts, Frankie Davis—the youngest daughter of the woman in question—called from Texas to offer more details. The story Michelle and Frankie told me was this: Leoma Patterson—Michelle's grandmother, and the mother of Frankie Davis and three other grown children—went missing from the town of Clinton, Tennessee (about twenty miles north of Knoxville), one night in October 1978. She left a bar with two men—a father and son—and was never seen again. Five months later some kids playing in the woods beside Norris Lake found part of a human skeleton. Animals had scattered or eaten many of the bones; the ones remaining on the wooded slope included a skull, some ribs, a couple of long bones, and part of the spine. The hair mat, which had sloughed off the skull as the scalp decayed, lay nearby as well, along with a tattered dress and a turquoise ring.

The remains had been examined by Dr. Cleland Blake, a medical examiner who served several East Tennessee counties. The bones had come from a middle-aged white woman, Dr. Blake concluded. Given that Leoma Patterson was the only missing person in the area who fit that profile, he tentatively identified them as hers. Frankie Davis—brought to the Campbell County courthouse by Special Agent David Ray, of the Tennessee Bureau of Investigation—identified the turquoise ring as Leoma's. It wasn't an airtight identification, but it was the best they could do, given the condition of the remains and the lack of a more definitive means of identification. Leoma Patterson had no fingerprints on file, but even if there *had* been prints on file, the remains had no fingerprints—no fingers at all, nor even hands—thanks to the ravages of time and the teeth of animals. Leoma Patterson also had no dental or medical X rays on file, nor were there other identifying skeletal features (healed fractures, for example) that her family could recall.

The mystery of how Leoma Patterson had ended up dead, dumped on a wooded hillside, remained unsolved for more than five years. Meanwhile, Dr. Blake had held onto the bones, against the possibility that the case might take a turn or investigators might get a break someday. And indeed, both of those events came to pass: In 1985 a great-nephew of Leoma Patterson's, Jimmy Ray Maggard—the younger of the men she'd left the bar with that night—confessed to killing her. They had quarreled over a drug deal, he said, and Maggard, nineteen at the time of the quarrel, had struck his fifty-six-year-old great-aunt with a tire iron. Maggard pleaded "no contest" to voluntary manslaughter, the case was closed, and the skeletal remains were finally buried in September 1985, beneath a heart-shaped granite headstone inscribed PATTERSON.

But Leoma's children—three daughters and a son—were never completely convinced by the medical examiner's report, never entirely sure that the identification was correct. Over the next two decades they would often wonder whether those really were their mother's bones in that hillside grave. Eventually, as their own children—Leoma's grandchildren—grew up, a new generation heard the story of Leoma's death, and heard the nagging doubts about the identification. By the time Leoma's granddaughter Michelle listened to me lecture in chemistry class, she'd heard the question countless times over the years: "Is it really her?"

Meanwhile, during the years since Leoma's disappearance, forensic science—especially techniques for human identification—had changed dramatically, thanks to a revolutionary new tool in to the forensic toolbox: DNA testing.

The DNA revolution had begun half a century before, in 1953, when two scientists at England's Cambridge University—James Watson and Francis Crick—solved a puzzle that had baffled geneticists and chemists for decades: What was the structure of the immense protein molecule that encoded every person's unique genetic "fingerprint," developmental blueprint, and biochemical operating instructions? The answer, which Watson and Crick deduced from micrographs taken by Rosalind Franklin as she fired X rays through crystallized specimens of the molecule, was elegant but surprisingly simple. That protein molecule, deoxyribonucleic acid—commonly known simply (and mercifully!) as DNA—was shaped like a microscopic ladder three billion rungs high, its uprights tightly twisted into the corkscrew double helix, which is now one of the most familiar shapes on Earth.

The breakthrough insight won Watson and Crick the Nobel Prize in 1962; it also paved the way, over the next three decades, for the development of analytical techniques capable of isolating any individual's unique genetic "fingerprint." But the path from laboratory capability to real-world applicability isn't always swift or smooth. Prosecutors learned this lesson the hard way in 1995—early in the forensic use of DNA—during the murder trial of O.J. Simpson. Despite DNA evidence strongly linking him to the murders of his wife, Nicole, and her friend Ron Goldman, Simpson was acquitted, in part because of perceived problems with the DNA evidence. Either the jurors didn't understand the science or they didn't trust the integrity of the police who handled the evidence. Or both.

Technology and public awareness advanced considerably after the O.J. trial, and by 2005, DNA testing had become routine, a staple of police departments nationwide—and a mainstay of television crime dramas. Show like *CSI* had made millions of people—including Michelle Atkins, a *CSI* fan—conversant with phrases like "blunt-force trauma" and "nuclear DNA" and "mitochondrial DNA." Hell, according to occasional snatches of dialogue on *CSI*, the show's main forensic genius even had a body farm much like mine, where he, too, studied postmortem human decomposition. So the show must be rigorously researched, incredibly accurate, and absolutely authentic, right?

Um, not exactly.

When I'm asked what I think about *CSI* and similar shows, I usually answer that they've done a great job of educating people about the importance of forensic science and the crucial role of combing crime scenes for evidence. They've also gotten countless young people interested in forensic careers; if I had a dollar for every student or parent who has come up to me in the past few years and asked how to become a forensic scientist, I'd be a rich man.

But the differences between TV forensics and real-life forensics are enormous. For one thing, TV detectives always solve the crime in an hour (actually, in forty minutes, if you subtract the commercials). The investigators almost never feel the need to ask questions, because they already seem to know everything. And on those rare occasions when they do feel the need to ask a question, they make one quick phone call and get an immediate answer. Similarly, they get DNA results almost instantly. Ask a hundred real-life homicide investigators how quickly they get DNA results, and I guarantee you not a one of them will tell you it's a swift process; in real life it generally takes weeks or even months, because certified forensic DNA labs are terribly backlogged.

In fact, there's such a gap between how forensic science is portrayed on TV and how it's practiced in real life (and real deaths) that beleaguered police officers and prosecutors have given the gap a name—the "*CSI* effect"—and almost anytime one of them utters the phrase, it's accompanied by head-shaking and eye-rolling and grumbling. On *CSI*, DNA testing is always swift and always infallible. But in my half century of forensic experience, swiftness and infallibility struck me as the exception; slowness and fallibility seemed far more common.

Therefore I felt some trepidation as I negotiated the winding road up and over and down the north side of Redoak Mountain to disinter a mountainside grave, and that trepidation wasn't entirely on account of the road's switchbacks and dizzying drops. A rural family, hailing from this hardscrabble area, had gotten the idea—partly from a Hollywood-style view of forensics—that science could now shed bright, infallible light on the fate of Leoma

Patterson, who had disappeared a quarter century before. I hoped they were right, but I couldn't be sure. I didn't want to give them false hope, and I didn't want to waste their time or money. This was not an affluent family by any means; in fact, I gathered that they had scrimped and saved for quite a while to pay for the exhumation, skeletal examination, and DNA testing. I hoped their hope was not misplaced, and that I could help resolve their nagging doubts once and for all.

I BREATHED A sigh of relief as our guide led us down the last rampart of Redoak Mountain and onto a short, straight stretch of road running along a valley floor. Not long after that, we turned off the pavement onto a steep gravel track angling up another hillside. A wheel-spinning, gravel-spitting quarter mile up the hill, we pulled off into a tiny cemetery and parked behind a handful of other pickups and SUVs, as well as a heavier truck hitched to a flatbed trailer. Leoma Patterson's three daughters stood waiting for us; so did three of her grand-daughters, along with sundry husbands and boyfriends and possibly a grandson or even a great-grandson or two. I didn't get a good handle on who the various quiet males were, because the women—two generations of Leoma's descendants—were clearly the ones running the show.

The flatbed trailer had hauled in a backhoe, and by the time Jon, Kate, and I got there, the grave had already been excavated and the lid of the coffin exposed, though it remained sealed. I shook hands with the family for a few minutes, then described what I was about to do. Next I clambered down into the grave to open the coffin and see what lay within.

The silver metal lid was slightly dented and somewhat rusted, but it opened easily. Inside, the fabric liner was soggy, stained with mold, and coming loose from the lid. There wasn't much else inside the coffin—a skull, perched on top of a wadded-up plastic bag, plus a few dozen other bones, some of them partially embedded in a thin layer of red mud that had seeped into the coffin over the past twenty years.

Above me, I heard the clicking of a camera shutter. Jon Jefferson had already begun his task of documenting everything. As I started removing the bones from the coffin, naming them one by one, Jon made a list and also marked each bone on a diagram of the human skeleton.

My other assistant, Kate Spradley, was a Ph.D. candidate in anthropology. I'd brought Kate because she was skilled in the use of a sophisticated computer program called ForDisc (short for *For*ensic *Dis*criminant Function Analysis), developed at UT by Richard Jantz and Steve Ousley. ForDisc was designed to compare an unknown specimen's skeletal measurements against measurements from thousands of known skeletons—specimens whose race, sex, and stature were known—and then calculate the race, sex, and stature of the unknown. Years before, when ForDisc was in its infancy, I'd been skeptical. Then, in 1991, I went head-to-head with ForDisc: Police had found a skeleton in a Tennessee creek and brought it to me to identify. The skeleton had some interesting, ambiguous features. In the end I decided that it was a white man's. ForDisc, on the other hand, was ninety-five percent sure it was a black man's . . . and ForDisc turned out to be right. Ever since that case, I'd had tremendous respect for ForDisc and its capabilities. It wasn't infallible, but neither was I, and two heads—mine and ForDisc's—were better than one, given the pressure to resolve the Patterson family's agonizing doubts once and for all.

To run the program, Kate had brought along a laptop computer and a three-dimensional digitizing probe—a high-tech tool resembling a small robotic arm—that she would use to take cranial measurements. After she plugged these measurements into ForDisc, the program would, in a matter of seconds, compare them to thousands of other skulls in its data base and then pronounce the skull to be white, black, Asian, or some ambiguous mixture of races. The computer and 3D digitizing probe would be powered by a portable generator, which we had hauled with us in the back of my truck. Kate's high-tech gear made for quite a study in visual contrasts, perched on the tailgate of a pickup, high on a backwoods hillside.

One of the first things I do with any case is assign it a number, which makes it easier to keep track of things over time. Years before, I'd adopted a system for numbering forensic cases: the current year, followed by the order in which the case had been opened. This was my first forensic case of 2005—making it case 05-01—but standing there beneath the PATTERSON headstone, in a grave ringed by a dozen or more descendants of Leoma Patterson, I couldn't help thinking of 05-01 as Leoma.

In identifying unknown skeletons, I always start with the "Big Four": sex, race, stature, and age. In this case, some of those traits would be easy to pin down; others, not so easy.

The skull could provide most of the answers, so I started there, lifting it carefully from its makeshift cushion. The family members edged closer as I examined the skull and began discussing its features and what they told me. The skull was small in size and smooth in texture—"gracile" is the technical term—as women's skulls generally are. It also lacked heavy brow ridges and prominent muscle markings—features common in male skulls. The upper edges of the eye orbits were sharp, the mastoid processes (behind the ears) were small, and the occipital bone, at the base of the skull, lacked the protruding bump that is common in men. If the pelvis had been present, its shape would have provided the final confirmation, but I didn't need it: The skull was clearly, classically female. One trait down, three to go.

Next came race. That, too, could be seen in the skull. The nasal opening was fairly narrow, with a well-defined sill at the base of the opening—features strongly characteristic of whites. Another white feature was a high-bridged nose. Those features, together with a mouth structure that was vertical ("orthognic") rather than angling forward ("prognathic," as in Negroid skulls), told me clearly that 05-01 was Caucasoid, or at least mostly so. Two down, two to go; so far, so good.

To estimate age, I looked at several features of the skull. First, I studied the teeth. As people approach adulthood, their third molars, or wisdom teeth, generally erupt around age eighteen years, although some people (more and more of us, in modern humans) lack third molars altogether. The mandible, or lower jaw, of 05-01 contained no third molars, but both of her upper third molars had erupted. That meant she was probably at least eighteen—information that was helpful, but not as specific as I needed. To narrow the age range, I looked next at the cranial sutures, the joints where the bones of the skull fuse together. At birth the cranial vault consists of seven separate bones, loosely joined to allow the head to grow easily. By age five the seven bones begin fusing at the sutures, zigzagging joints that interlock rather like the teeth of zippers. Around age thirty the sutures generally begin to ossify—to fill in with bone and smooth out—and eventually the sutures may become completely invisible, or obliterated. Cranial sutures don't allow age estimates to be made with pinpoint accuracy by any means, but they can often narrow the range to within a decade or so, which can help investigators considerably: It's much easier to search records for missing females aged forty to fifty than for females aged twenty to eighty.

The first of the cranial sutures to close—beginning around age twenty—is the basilor suture, located at the base of the skull, between the occipital bone and the sphenoid. Holding the skull from the coffin upside-down in my hand, I showed the family the basilor suture. "It's fully closed," I told them, explaining that meant she was at least twenty-five. Next I studied the sutures along the top and sides of the skull—the coronal, sagittal, and lambdoidal sutures—and saw that they were nearly obliterated. The zigzag lines that appear so stark in skulls of twenty-five-year-olds were only faintly discernible in this skull; in fact, they looked almost as if they'd been filled with spackling compound and then sanded smooth.

Mentally comparing these sutures to the thousands of others I'd studied over the decades, I estimated the woman's age to be at least forty. I didn't think she could be elderly, though, because she lacked the signs of skeletal wear and tear that characterize old people. For one thing, she still had all of her teeth—quite an accomplishment for someone growing up in the 1930s and '40s, especially in the hills of East Tennessee! For another, she showed few signs of osteoarthritic lipping, the buildup of jagged ridges of bony material along the edges of vertebrae and other joint surfaces. Think of lipping as the skeleton's version of the mineral deposits that gradually build up and clog a house's water lines; not surprisingly, the bony fringe is a major contributor to the aching joints that plague the elderly. Whenever I show slides of severe osteoarthritic lipping, someone always asks if there's any way to prevent it. "Of *course* there is," I exclaim. "Die young!" Well, this woman hadn't died young, but she did seem to have died in middle age, rather than old age.

But when I took a closer look at her teeth, I was surprised to see how worn they were, at least on their occlusal surfaces—the edges that do the work of biting food and fingernails and pencils. When I fitted the mandible into its normal position beneath the skull, I saw why. Rather than having a slight overbite, as most white people do, this woman had an edge-to-edge bite; her top and bottom teeth lined up and made contact all the way around the arch of her mouth. During decades of chewing, those occlusal surfaces had done more than just tear meat and grind corn; they had also ground away at each other, wearing down the enamel a tiny bit each year. Native Americans tend to have edge-to-edge bites, so I asked the family members whether they knew if Leoma had any Indian ancestors. Yes, someone told me, she was supposedly part Cherokee—one-quarter or one-eighth. The edge-to-edge bite and occlusal wear also, therefore, suggested that this was indeed Leoma.

But one other feature of the skull seemed to contradict that notion. The feature—or was it a *non*-feature?— was skull trauma, which was notably absent. Jimmy Ray Maggard had told investigators that he'd killed Leoma with a tire iron, yet this skull was intact and undamaged. In most cases, killing a person with a tire iron leaves a skull fracture so sharp and recognizable that it's considered a "signature fracture." Other implements that leave signature fractures include claw hammers and golf putters. (I know this about putters thanks to a skull in our forensic collection: a skull into which a neat rectangular hole has been punched—a hole that the broken putter recovered from the crime scene fits into as precisely as a key fits a lock.) Still, there was a possible explanation for the inconsistency. According to several family members, Jimmy Ray Maggard had a tendency to tell outlandish lies, so it was possible that the tire iron was an after-the-fact embellishment.

Most of the woman's postcranial bones—the bones below the skull—were missing, and of the ones that were present, many had been chewed by carnivores. Only two long bones remained, the right humerus (upper arm) and the left femur (thigh), and neither of these was complete. The distal (elbow) end of the humerus had been chewed off, and both ends of the femur were gone—testament to how irresistible dogs find the marrow that lies within the ends of the long bones. Dr. Blake's crime-scene report had indicated that an intact left humerus had been found at the scene, but his autopsy report didn't list this bone in his inventory of skeletal elements. I hoped this inconsistency would be resolved by the presence of a left humerus in the coffin, but there was none. Lacking a complete long bone, I would be unable to calculate the woman's stature.

Fifteen vertebrae were strung on a rotting loop of twine, forming a sort of macabre necklace; it had been created not as ghoulish jewelry but simply as a practical way to keep the vertebrae together. The first four cervical (neck) vertebrae were missing, and so was the twelfth thoracic (chest) vertebra, but eleven of the twelve thoracic vertebrae were present. That didn't surprise me, as the crime-scene report had indicated that much of the upper torso remained, including much of the right rib cage, which had still been covered with leathery skin at the time the remains were found near the lakeshore. Apart from the lack of osteoarthritic lipping, the vertebrae didn't give me any significant information.

The ribs showed extensive signs of carnivore activity. Five of the left ribs were missing altogether, and only the head and neck regions—stubs, in other words—remained of the other seven. The right ribs had fared considerably better; surprisingly, in fact, all twelve of them were still there. Two adjacent right ribs—either ribs six and seven, or seven and eight (it's hard to distinguish between the middle ribs, especially if they're incomplete) showed signs of trauma that was *not* caused by carnivores. At midshaft, both ribs exhibited a hinge fracture, suggesting that a blow to the right chest caused these two ribs to fold inward. Could that blow have caused her death? Perhaps, but it seemed unlikely, and without soft tissue, there was no way to know.

Other potential, unknowable causes of death (if she hadn't been bludgeoned with a tire iron) included shooting, stabbing, and—twice as common in women's murders as in men's—strangulation. I had noticed, in reading Dr. Blake's crime-scene report, that he had recovered the hyoid, and I was looking forward to examining it. The hyoid is a delicate, U-shaped bone from the neck. Free floating—that is, unattached to any other bones—the hyoid supports the muscles of the tongue. I was especially eager to see the hyoid because it's sometimes a telltale bone: If it's fractured or crushed, it points strongly to strangulation. I searched the muddy coffin carefully, but to my dismay the hyoid was nowhere to be found. *Too bad*, I thought, because the other woman Jimmy Ray Maggard had confessed to killing—a Georgia woman—had been strangled.

Still, I wasn't there to investigate the cause or manner of death; I was there to obtain DNA samples, to help answer the question that had haunted Leoma Patterson's family all these years: Was this really her?

I had come equipped to take two kinds of samples—teeth, and cross-sections of long bones—to raise the odds that I'd get intact DNA so many years after death. When exposed to the elements, to bacteria, and perhaps to the body's own decomposition processes and products, DNA gradually degrades. That meant the best places to look for good DNA samples would be within molars or from the shaft of the long bones—the places where there was the most protection against chemicals and microbes. Using a pair of pliers, I carefully extracted two teeth from the jaws: the right first molar from the mandible, and the left second molar from the maxilla. I placed each tooth in a sterile plastic sample vial, which I labeled and gave to Frankie Davis. Frankie, Leoma's youngest daughter, seemed to be the most sophisticated of Leoma's children, and she had offered to find a DNA lab to analyze the samples and compare their genetic material with cheek-swab samples from herself and her sister Pearl.

Extracting the teeth was easy. Getting DNA samples from the long bones was slightly harder. For that, I had brought a Stryker autopsy saw, along with a hundred-foot extension cord. We plugged the cord into the portable generator I had brought, and while Jon steadied the bones—bracing them on the grave's heart-shaped headstone—I cut thin cross-sections from the shafts of the right humerus and the left femur. I sealed each of these in a sample vial as well and handed them off to Frankie.

As soon as I finished collecting the DNA samples, Kate began taking skull measurements at her makeshift workstation, the tip of the probe darting swiftly from landmark to landmark: the bridge of the nose, the centers of the eye orbits, the tip of the chin, the crown of the head, the outer corners of the cheekbones, and so on. Unlike traditional calipers, which have to be carefully aligned on two points simultaneously to take measurements, the probe can simply be touched at each point individually; somehow (don't ask me how) the computer assigns relative 3D coordinates to every point the tip contacts. As a result, the software can easily calculate, say, the distance between the eye orbits, or the width and height of the nasal opening, or the degree of prognathism in the mouth structure, and so on. Back when I was a graduate student working at the Smithsonian Institution, taking thousands of skull measurements every week, I could make three to four cranial measurements per minute—provided I had someone with me to record the measurements as I called them out. Because the digitizer's sharp tip is easier to position than the ends of calipers (which can slide off the skull's domed surfaces, or prove difficult to fit into small openings), Kate could take a measurement every second—single-handedly, and with greater precision.

There in the cemetery, Kate ran a preliminary analysis of the measurements, which she explained to the family. Like me, ForDisc had a high degree of certainty that the skull was a white woman's, possibly with a bit of Native American ancestry mixed in. She and I spent a while answering more questions from Leoma Patterson's descendants. Eventually, though, people ran out of questions. Clambering back into the open grave, I laid the bones back in the coffin, closed the lid, and resealed it. Someone reached down to help me out of the grave, and the backhoe operator fired up the machine and filled the hole. Kate, Jon, and I said our good-byes, and I backed out of the cemetery, eased down the gravel road to the valley floor, and then snaked up and over Redoak Mountain once more, back home to Knoxville.

A few days later I wrote my report. In the summary section, I concluded this: "There is nothing in the skeletal material that I looked at that would not be expected in the skeleton of a fifty-year-old white female with American Indian ancestry."

FOUR MONTHS PASSED, and the case slipped from the forefront of my mind to the background. Then, in December, Frankie Davis called from Texas to say that she had just received a report from GenQuest, the DNA laboratory to which she'd sent the samples. GenQuest had compared the skeletal material's DNA to the cheek swabs from Frankie and her sister Pearl. After examining ten different regions of DNA from the various samples, the laboratory concluded that the woman in the grave was unrelated to Frankie and Pearl. I felt sure I'd misunderstood, so I asked her to repeat it, and she did. I had not misunderstood. According to the lab, the woman in the grave was not Frankie's mother. She was not who the granite headstone proclaimed her to be. In short, she was not, and never had been, Leoma Patterson.

The news stunned me. If that wasn't Leoma Patterson buried in the family cemetery, then who the hell was it? And where the hell was the real Leoma?

For photos from Chapter 1, please visit: http://www.jeffersonbass.com/books/identitycrisis/ch1.html

Chapter Two

Putting a Face on the Dead

I HAD THOUGHT, when we reburied the coffin beneath the heart-shaped headstone on that hot August afternoon, that my work on case 05-01 was finished. Instead, it turned out, my work was just beginning. On December 2, 2006—fifteen months after my first unnerving journey over Redoak Mountain—I retraced the switchbacks, this time in a Honda minivan I had recently bought. This time I didn't get lost, and I didn't get scared. This time the hairpin curves and sharp drop-offs of the road concerned me far less than the baffling turn the case had taken—a case I'd thought would be so simple. Like Leoma Patterson's family, I was now confused and frustrated. And I was determined to discover the identity of the mysterious woman whose bones had been misidentified a quarter century before and then subsequently buried in a family cemetery—a cemetery where they did not belong.

Once more surrounded by Leoma Patterson's descendants, I clambered down into the raw, reopened wound in the earth. After a prayer by one of Leoma's grandsons—a prayer imploring God to right the wrongs that had been inflicted on the family by the miscarriage of justice, and to help us find the truth at last—I opened the battered coffin and removed the bones again. This time, though, we spent no time on high-tech 3-D measurements or detailed explanations of skeletal features. In very short order the coffin was reburied—again—but this time it was empty, except for the thin layer of mud inside. This time the bones were coming with me back to Knoxville. This time 05-01 was the center of a new investigation.

When Frankie Davis had dropped the GenQuest bombshell on me, I quickly relayed the news to Paul Phillips, the district attorney in Campbell County, where the bones had been found in 1979. Paul had opened a new investigation, and he'd assigned a Tennessee Bureau of Investigation agent, Steve Vinsant, to the case. But whose case would Vinsant be investigating? Whose killer might Paul Phillips someday prosecute? Who was this middle-aged Jane Doe? In my files, the case number remained the same—to me, she was still 05-01. But I could no longer think of her as "Leoma." From now on she would become "Not-Leoma." I hoped to discover Not-Leoma's true identity in Knoxville—specifically, in UT's immense temple to college football, Neyland Stadium.

By this point I had spent much of the past thirty-five

years in Neyland Stadium. *Deep* in Neyland Stadium. Strangely, the Anthropology Department is tucked beneath the stadium's south and east grandstands, in a dingy bridge wedge of offices, labs, and classrooms. Even more strangely, the grimy, girder-shadowed location was of my own choosing. Allow me to explain. Back in 1971, I was hired to head UT's Anthropology Department, and to expand it by adding a master's degree and doctoral program. At the time, the department was housed in McClung Museum, and it was immediately apparent that there was nowhere in McClung for us to grow. We had to go. I was offered a choice: a couple of dilapidated old houses UT owned, or the outmoded athletic dorm, Stadium Hall. I chose Stadium Hall; it was grimy and gloomy, but it was big and it was damned sturdy. In any case, the quarters were only temporary, I was assured. Right.

A brief historical, scientific, and architectural digression seems in order here. The world's first controlled nuclear chain reaction, in 1942, occurred in a rustic laboratory wedged beneath the grandstands of the University of Chicago's football stadium. So I understand, and even appreciate, that there *is* a precedent for substadium brilliance. Still, it strikes me as ironic that at UT anthropologists are housed in the same sort of quarters where potential *atomic catastrophes* were quarantined at Chicago.

Even deeper within Stadium Hall from the Anthropology Department's offices and classrooms—another flight down, wedged between two staircases—is a large room with one wall of windows, two computers, a pair of desks, half a dozen or so long metal tables, and tens of thousands of bones. The bones are Native American skeletons that teams of students and I excavated in South Dakota back in the 1950s and '60s, when new dams on the Missouri River began creating vast lakes and flooding abandoned Arikara Indian villages. This basement room is the osteology laboratory—"the bone lab," everyone in the department calls it—and it's here that the skulls and skeletons of the unknown and the murdered get measured, examined, discussed, and often identified by the Anthropology Department's forensic faculty and graduate students. If determination, knowledge, and luck combined in sufficient measure, the bone lab was the place where we might discover the true identity of Not-Leoma. Of 05-01.

To identify an unknown victim through DNA fingerprinting—or conventional fingerprinting, for that matter—you must compare at least two sets of fingerprints: the unknown's prints, and the prints of a known person that could potentially be a match. In the case of the bones we had twice exhumed from beneath Leoma Patterson's headstone, we had no idea where to begin looking for a potential match. No other middle-aged East Tennessee white woman had gone missing and *stayed* missing in the late 1970s. We expanded the search area, combing the missing-persons' data base of the National Crime Information Center for a possible match regionally and nationally, but we found no likely matches there either. That left us with one last, long-shot hope that might lead to an identification. In football, a desperate, unlikely effort to snatch victory from defeat at the last instant is called a Hail Mary pass: You fling the ball far down the field and pray. We were about to try the forensic equivalent of a Hail Mary pass. A young woman named Joanna Hughes was going to throw it for us.

ON THE FAR side of the bone lab—opposite the single steel door, which is locked whenever the lab is unoccupied (by the living)—sits a small wooden table that looks as if it's been teleported from a studio in the Art Department, complete with art supplies: clay, palette knives, even a striking bust or two, sculpted in soft, lustrous gray clay. The sculptures are the handiwork of Joanna Hughes, a talented artist who is, as far as I know, the only person in the United States with a degree in forensic art. She earned it at UT, creating the program herself by combining traditional art and sculpture classes with extensive study of anatomy and anthropology. Joanna's rare but valuable talent—her career and her calling—is to restore the faces of the unknown dead, working only from bare skulls, raw clay, extensive knowledge, artist's intuition, and tenacious faith in long-shot, late-in-the-game breakthroughs. Joanna puts faces on the dead in hopes of coaxing a fading memory of a long-gone visage up to the surface of someone's consciousness.

In most cases, by the time an artist like Joanna starts layering clay on bone, years have passed since the final smile or tear or moan or plea. Not surprisingly, the odds are slim that an identification will ever result from an artist's educated guess at the face that a skull once wore, years before. Those long odds made Joanna's track record rather remarkable: Of the nine reconstructions she had done in the prior five years, three led people who saw photographs of the sculptures to come forward and say, "I know who that is." Or rather, "I know who that was."

The DNA report from GenQuest, which concluded that the DNA from the bones and teeth in Leoma Patterson's grave did not match the DNA from two of Leoma's daughters, had opened a can of worms that was frustrating but also fascinating. Instead of an old, closed case, we now had two baffling, open cases, and two maddening questions: Where was the real Leoma Patterson? And whose bones had we removed from that long mismarked grave? With luck and a skillful reconstruction—plus wide distribution of the reconstruction's photograph in newspapers and on the TV news—Joanna might help us answer the second question.

For some people, forensic work is just a job. Joanna's work arose from a mysterious calling—one that later came to look like destiny. Her story is astonishing, heartrending, and ultimately inspiring. She grew up in a small town in south Alabama, Monroeville, population 6,690. Monroeville's main claim to fame thus far is that it's the hometown of novelist Harper Lee, as well as the fictionalized setting of her classic novel, *To Kill a Mockingbird*. Monroeville was also the boyhood home of Lee's friend Truman Capote, author of *In Cold Blood*. With its tree-lined streets, columned antebellum houses, elegant 1903 courthouse, and quaint general store ("O.B. Finklea Store

No. 2, Dealing in What You Want—Dry Goods, Shoes, Groceries, School Supplies, Drugs, Hardware, Fertilizer"), Monroeville could, at least in places, easily pass for the 1930s town that Harper Lee described in her novel: "an old town . . . a tired old town when I first knew it." By the time Joanna came along, it was even older and more tired.

Joanna never knew her biological parents; as a baby, she was adopted by Tim and Nancy Jones. Tim was the quintessential small-town doctor. He worked long hours, treated wealthy and poor patients alike (well, not *exactly* alike, because he didn't charge the poor ones as much), and held a place of both respect and affection in the community. His wife, Nancy, was a one-woman civic institution. She made home-cooked meals for new residents, delivered toys to their children, and baked bread for old folks. A deeply religious woman, she trusted in the transforming power of faith.

At age twelve, Joanna startled her parents by announcing, "Someday I want to put faces on skulls." Looking back, she has no idea where she got such a notion; she only recalls that the vision was quite specific and crystal clear. What was less clear was how to turn it into reality, and into a paying job. So when it came time for college, she opted for something a bit (though perhaps only a bit) more practical: a degree in film. At age nineteen, after her first year of film school at Florida State, she landed a job as a production assistant on a Steven Segal film, *Under Siege*. It was her first taste of Hollywood, and it soured her on the idea. She was the only young woman on the set, surrounded by "a bunch of horny guys," in her words. She completed her degree in film anyhow, but after graduating, she took a job as a secretary to earn a living and then began to do what she'd wanted to do in the first place. She enrolled in art and anthropology classes at UT, and she persuaded university officials to let her design a program custom-tailored to give her a bachelor's degree in forensic art.

In the fall of 2000, Joanna began doing facial reconstructions for homicide investigators on a volunteer basis. She was finally pursuing her dream, but not for the money, because so far it was strictly pro bono. The following year, in a National Geographic documentary about the Body Farm, she was challenged to put her art to the acid test, before an audience of millions. The documentary producer—none other than Jon Jefferson— handed Joanna the bare skull of a man she had never seen and said, in effect, "Let's see how good you are." Jon knew what the dead man had looked like—he and his crew had spent two months filming the corpse's decomposition— but Joanna didn't. All she knew was that the man was elderly and of Greek ancestry.

Joanna labored for two weeks to reconstruct the dead man's face. When she finally finished, Jon handed her a sealed envelope containing a large photo. Nervously she opened the envelope and slowly slid it out. Upon seeing the face in the picture—a face that looked remarkably like the clay reconstruction on the table in front of her— Joanna beamed and said just two words: "It's *him*!"

It was impressive, yes, but it was only a demonstration.

In the fall of 2002, Joanna got her first paying commissions to do reconstructions in forensic cases: the skulls of two murdered African American women, their decomposed bodies found together in the woods near Petersburg, Virginia, earlier that year. Two years after police and the Doe Network published the photos of Joanna's reconstructions, the first of the two women was identified; in 2006, so was the second.

Joanna found putting faces on the dead to be intellectually and artistically satisfying, but not financially rewarding; each reconstruction took two weeks of concentrated effort, and she wasn't being paid. She considered going back to film production. Then one morning her life took a shattering turn.

Joanna wasn't the only child who had been adopted by Tim and Nancy Jones. When she was a toddler, her parents had adopted another baby, a boy they named Timothy Jason. Joanna took an instant dislike to Jason, her new brother. "I told them there was something wrong with him," she recalls. "I told them to take him back and get a different one." At some level, she had sensed something frightening in Jason, and over the next two decades her intuition would be borne out time and time again. Jason was an angry child, prone to violent outbursts—a source of bewilderment and sadness to parents who were kindhearted and gentle by nature. By his teens, Jason was using drugs and failing high school. He dropped out, though he later passed the GED (General Equivalency Diploma) test. He spent years drifting between short-lived jobs, drug-rehab clinics, and occasional stints in jail. Joanna was a frequent caller to the Monroeville police; her calls tended to go something like this: "Hey, this is Joanna; could you come to my parents' house? Jason stole something again." But Tim and Nancy Jones refused to give up on him, and kept trying to help him turn his life around.

They kept trying until January 29, 2004. Early that morning, as Dr. Jones was in his driveway, about to leave for his clinic, Jason bludgeoned his adoptive father to death with a steel pipe. Then he went inside and killed his mother as she lay in bed. Nancy Jones was bludgeoned so savagely, according to the Mobile *Press-Register*, that "parts of her jaw and teeth were found scattered across the room." After the murders, Jason took money from Nancy's purse, went out and bought crack cocaine, and returned to the house to smoke it. Then he set out in Nancy's car for a girlfriend's house in north Alabama. The car and the trip were apparently the motive for Jason's murderous rage: His parents had recently taken back an SUV they had loaned him, and they refused to let him borrow it for the trip north, because the trip would have violated the terms of his probation. So he retaliated by slaughtering them.

At his murder trial in Birmingham, Joanna testified for the prosecution about her brother's chronic drug use and violent temper. She felt relief when he was convicted, and felt a grim sense of justice when he received the death sentence. She even felt icy satisfaction in September 2006 when Jason—on death row at Holman Prison in Atmore, Alabama—slashed his own neck with a makeshift blade and bled to death in his cell. In a final twist of fate, Joanna—Jason's next of kin—was asked by the prison authorities what they should do with his body. Her answer stemmed partly from a desire to contribute to science and partly from an impulse for posthumous revenge. "Send him to the Body Farm," she said, and they did.

The tragedy in her own family cemented Joanna's commitment to a forensic career. Although she can't bring back her parents, she can potentially help bring killers to justice, by restoring faces and identities to unknown murder victims—victims like 05-01, Not-Leoma, whose body was dumped in the woods by Norris Lake more than twenty-five years before.

JOANNA BEGAN THE reconstruction by simply looking at the skull, noticing its general shape and unique features. Two characteristics caught her trained eye right away, much as they had caught mine when I had first lifted the skull from the coffin: the high, wide bridge of the nose, and the broad, strong chin. Most women have sharp chins, but not this one.

Then came a time-consuming tedious step in the process: cutting tissue-depth markers and gluing them onto the skull, to guide her as she applied the clay and began to shape its contours. To understand the importance of this step, take a moment to do a simple experiment: Press the center of your forehead with a fingertip; there's not much tissue between your finger and the frontal bone, is there? Now feel your cheekbones: There's more flesh atop those, but still not a lot. Now drop down between your cheek-bones and lower jaw: There's a lot of flesh and muscle there, including the muscles you use to smile, and those you use to chew. The variations in soft-tissue thickness complicate the work of facial reconstruction artists. Fortunately, they have a large body of research—or perhaps I should say "many bodies of research"—to guide them.

Starting in the late 1800s, a series of German anatomists took careful measurements, from thousands of cadavers, of the thickness of tissue at numerous points on the head, especially the face. Their technique was simple but ingenious: They inserted a needle through a small cork, until the tip of the needle was just starting to emerge through the other side of the cork. Then they rested the cork on the skin of a cadaver's face and pushed the needle farther, through the skin, until the needle hit bone at each of dozens of landmarks on the cadaver's skull—landmarks with catchy names like "nasion" (where the nasal bone joins the skull), "pogonion" (the most anterior, or forward, point at the center of the chin), "bregma" (the top of the skull, where the frontal bone joins the two parietal bones), "ectomolare" (the point in the upper jaw where the second molar meets the bone), and "glabella" (the most forward point in the forehead, between the ridges above the eyes). After each insertion, they carefully withdrew the needle and measured how far the tip extended beyond the cork, which had stopped at the surface of the skin. Being Germans, the needle-wielding anatomists were meticulous and thorough, gathering data on men, women, and children of various ages and races. Their simple, ingenious measurement technique did have one built-in flaw: The slight bit of friction between the needle and the cork caused the cork to depress the skin slightly as the needle was inserted, so their measurements consistently understated the thickness of the tissue. Still, it was pioneering research, and its one shortcoming can now be avoided by measuring tissue thickness with magnetic resonance imaging, by scanning cadavers in MRI machines.

For 05-01, Joanna would use the tissue-depth data for adult women of European descent. Her first step would be to measure and cut tissue-depth markers from long, cylindrical pencil erasers, the type used in Click-It mechanical erasers by artists and architects. The markers ranged from a mere 2.75 millimeters long (just over a tenth of an inch)—for the thin tissue at the end of the nasal bone—to seventeen mm (two-thirds of an inch), for the occlusal line, the fleshy region where the upper and lower teeth meet. Next, she would glue these markers to seventeen landmarks on the skull. The process of cutting and gluing the markers would take hours, but without the markers, she'd be working blindly, with no way to tell how thick the clay was getting as she molded and shaped it atop the bone.

The traditional way of doing a forensic facial reconstruction—the clay way; Joanna's way—is a tedious, exacting process. It's the sort of task that seems like a prime candidate for computerizing. You might think it would be a simple matter to program a computer to add a face to the image of a skull; after all, isn't it just a matter of adding so many millimeters of padding here, so many millimeters there, smoothing the transitions, tinting the surface—oh, and plugging in eyes, lips, and a nose? Well, it might be straightforward to program a computer to do those things, but apparently it's devilishly difficult to program the computer to do it *well*—to breathe life into the image. Over the years, I had heard about, read about, and seen the results of various attempts to computerize facial reconstruction, and the images resulting from these attempts always looked to me like Claymation figures or lifeless masks. Science can provide the foundation for a good facial reconstruction, but—at least from what I'd seen so far—it takes art to bring it to life.

But I was willing to be persuaded otherwise. I'd heard about an ambitious new attempt to computerize facial reconstruction—this one by the FBI—and it seemed worth a shot. One of the Anthropology Department's graduate students, Diana Moyers, was working as a visiting scientist at the FBI laboratory in Quantico, Virginia. Diana's project there was to test and help refine an experimental computer program called "ReFace"—facial-reconstruction software whose full name is Reality Enhanced Facial Approximation by Computational Estimation. (You can see why they shortened that to ReFace!) Developed for the FBI by GE Global Research, ReFace starts with skulls—three-dimensional CT scans of skulls, to be precise—and uses tissue-depth data and algorithms (fancy mathematical formulas) to do virtually, on a computer screen, what an artist like Joanna does with clay on a skull. Being a computer program, ReFace could work more objectively than Joanna; it could also work far faster, completing a 3-D rendering in a matter of hours rather than weeks. Such blazing speed gives ReFace great potential for helping identify victims found in mass graves, as in Iraq. If it had been available at the time, it also could have helped enormously in the wake of Hurricane Katrina, when forensic teams struggled for months to identify hundreds of badly decomposed bodies. If computerized facial reconstructions of those storm victims could have been easily produced and widely distributed, the task of restoring names to Katrina's dead might have moved forward far faster.

Jon and I contacted Diana and her FBI supervisor, Phil Williams, to ask for help. To our delight, they agreed to put ReFace to the challenge, even though the software was still in the experimental stage. All they needed was a CT scan of the skull. That was easy to arrange, since one of the anthropology graduate students, Megan Moore, had recently scanned every skeleton—more than six hundred—in the donated skeletal collection; one more skull was just a final drop in the bucket. Megan and Todd Malone, a CT technician in the Radiology Department at UT Medical Center, ran skull 05-01 through the scanner, faceup, in a box that was packed with foam peanuts to hold it steady. Megan FedExed the scans to Quantico, where Diana and Phil Williams ran them through the experimental software.

It was with high hopes, shortly after the scan, that I studied the computer screen showing the features ReFace had overlaid, with mathematical precision, atop the CT scan of Maybe-Leoma's skull. Surely this image, I thought—the fruit of several years of collaboration by computer scientists, forensic artists, and anthropologists—would clearly settle the question of 05-01's identity: Was she Leoma or was she Not-Leoma? Instead, the image merely amplified the question. The flesh-toned image on the screen—eyes closed, the features impassive—could have been a department-store mannequin, or a sphinx. There was nothing in the image, no matter how I rotated it in three dimensions, that said, "I am Leoma." Nor was there anything that said, "I am not Leoma." To borrow Winston Churchill's famous description of Russia, the masklike face on the screen was "a riddle wrapped in a mystery inside an enigma." Between the scan, the software, and the tissue-depth data that the software merged with the scan, it was all very sophisticated and high-tech. But it was still a damn puzzle, and the identity of the dead woman remained as elusive as ever.

I HAD TAKEN the skull to Joanna in December 2006, shortly before Christmas. An eternity later—in mid-January, actually, only three weeks later—she was ready to show me what she had done. It wasn't quite finished, she said, but it was getting close.

When I walked into the bone lab and saw her handiwork, you could have knocked me over with a feather. Joanna's reconstruction—based solely on the shape of the skull, the scientific data on tissue depth, and the information I had given her about the age and race of 05-01—bore an uncanny resemblance to Leoma's daughter Barbara. It also looked strikingly like a photo I'd been given of Leoma Patterson herself, taken when Leoma was in her twenties—before her face puffed and sagged, as other, later photos showed it had. The more I compared the clay reconstruction with the photo of Leoma, the more remarkable the resemblance seemed—and the more troubling. What were the odds that two women, both in their fifties, would go missing in the same area of East Tennessee . . . and that they would look enough alike to be able to pass for sisters?

I had promised to send pictures of the reconstruction to various members of the Patterson family as soon as they were available, and Joanna made it easy, by taking digital photos and e-mailing them to Jon, who is more computer savvy than I am. Jon, in turn, relayed them to three of Leoma's children—Frankie Davis, Barbara Atkins, and Ronnie Patterson (who appears in Leoma's youthful photo as a small and solemn-faced baby sitting on her lap)—and to two of her grandchildren, J.R. Roach (the son of Leoma's oldest daughter) and Nancy Albert (the daughter of her youngest daughter).

Nancy was the first to weigh in. "I was shocked," she said. "As soon as I saw it, I thought, 'That looks like my grandmother.'"

J.R. Roach—at forty-four, the oldest of the grandchildren—was next. He was far less positive. "I don't see a whole lot of resemblance," he said.

Barbara was the daughter Jon and I thought the reconstruction most resembled—"dead ringer" (no pun intended) was the phrase Jon used. So I was startled when Barbara said flatly, "That's not her." A moment later her brother Ronnie got on the phone. "That don't look anything like our mama," he said.

The strong difference of opinion took me by surprise, though maybe it shouldn't have. Memory is a tricky and unreliable thing, facial reconstructions are approximations at best, and resemblance is in the eye of the beholder. What's more, there had always been disagreement within the family about the original identification: Frankie Davis and her daughter Nancy were convinced that a turquoise ring found with the remains in 1979 had belonged to Leoma ("I saw a picture of my grandmother wearing that ring," Nancy said), but other descendants insisted that the only ring Leoma ever wore was one that her brother had fashioned for her from a silver dime.

Within the Anthropology Department, opinion was more consistent, though less passionate. Kate Spradley, the Ph.D. student who had taken the digital measurements of the skull, took one look and said, "That looks like the family members." Lee Jantz, who runs the bone lab, studied the reconstruction closely, comparing it to our best photo of Leoma's face. Lee—who knew about the GenQuest report—raised her eyebrows and said slowly, "That is *very* interesting!" Then, after a pause, she added, "But the DNA says it's not her. So it's not her. Can't be."

She had a point. The DNA test had been conclusive. Or had it? Slowly—faintly at first—alarm bells began to ring in the back of my mind. I dug out the GenQuest report and pored over it. "Based on the above data," it read, "the profiles obtained for Pearl Smith and Frankie Davis have identical sequence variations. It is likely that Pearl Smith and Frankie Davis are from the same maternal lineage." In other words, these two sisters had the same mother: not exactly a news flash. The bombshell was what followed: "In comparison to the bone sample," the report went on, "Pearl Smith and Frankie Davis can be excluded as having the same maternal lineage." There: that part sounded conclusive, and *exclusive*. The woman whose teeth and bones I had sampled was not the mother of Pearl and Frankie; was not, in other words, Leoma Patterson. It was that line in the report that had led us to stop referring to 05-01 as "Leoma" and to start calling her "Not-Leoma."

But as I read and reread the report, other sentences slowly caught my eye, and my attention. "The sample tested was a piece of bone." Why just a piece of bone, rather than either of the teeth? Then—in a small footnote to the data about the bone sample, I read this: "The presence of multiple probes prohibited the identification of a single mitochondrial DNA profile. Possible explanations include that the sample is degraded. It is not recommended to use this information for sole identification or comparison purposes." Finally, at the bottom of the report, was this note: "Direct sequencing of the HVI and HVII regions is recommended to determine conclusive results."

The more I studied the GenQuest report, the more contradictory it seemed. On the one hand, it seemed to say that the woman in the grave was not Leoma; on the other, it seemed to say that the conclusion wasn't particularly conclusive, wasn't necessarily reliable. DNA isn't my field of expertise; hand me a shattered skull or a knife-marked rib, and I feel confident that I'll see what's forensically important. Talk to me about nucleotides or restriction enzymes or sequencing, and it soon starts sounding like Greek to me. I decided to consult an expert. Luckily, one was right at hand, unpacking her books and files.

At about the same time Joanna Hughes began gluing depth markers onto our baffling skull, Graciela Cabana, Ph.D., had arrived in Knoxville and begun settling into her new office beneath Neyland Stadium. Graciela was an important addition to the Anthropology Department: a molecular anthropologist, one specializing in DNA. I desperately hoped she could give me an objective reading and clear explanation of the GenQuest report.

We met in the office of Pam Poe, an administrative assistant in the Anthropology Department, because Graciela's office didn't yet have a desk, a phone, or even a chair. Jon and I were joined by Lee Jantz, who oversees the Body Farm's donation program, in addition to the bone lab, as well as Lee's husband, Richard Jantz (co-developer of ForDisc), head of the department's Forensic Anthropology Center. This case was a head-scratcher, and Lee and Richard were getting caught up in it, too.

Graciela began slowly, cautiously. "I can't quite tell," she said, "what GenQuest did and how they did it." She had studied the report itself; what's more, she had called the lab and—without indicating exactly why she was calling—asked the lab's chief scientist some questions about their procedures and protocols. The fact that she remained puzzled even after talking to the senior scientist? That didn't sound good. "For one thing, I can't tell which sample they used," she explained. "It appears they used only one bone sample, which surprises me. With old material like this, teeth are almost always a better source of DNA than bones." I had known this—the hard shell of enamel on the tooth encapsulates the DNA and protects it from degradation better than bone; that was why I'd included two teeth in the set of four samples. Next, Graciela focused on the lab's inability to establish a single profile for the dead woman. As she went on, her caution gradually dropped away. "This report is not just inconclusive," she finally said, to the astonishment of us all, "it's completely worthless. The only thing you can tell from this report," she concluded, "is that the sample was degraded or contaminated. Maybe both."

The air in the office nearly crackled with electricity. Suddenly everything—every possibility—was up for grabs again. Originally, the woman in the grave had been known as Leoma; after the GenQuest analysis, we had taken to calling her Not-Leoma. Now, in the space of a few minutes, she had suddenly become Maybe-Leoma. I couldn't help thinking of *CSI* again. On television, DNA testing provides an instant, ironclad answer; in this real-life case, though, DNA testing had merely—and hugely—muddied the water: It appeared to provide an answer, but in fact all it did was raise a whole new set of questions—not just for me, but for Leoma Patterson's relatives. The family had mistrusted the original investigation's findings, which is why they contacted me in the first place, but I couldn't imagine that this new twist would do much to restore their confidence. And it certainly wasn't likely to ease their pain.

Now what? The answer seemed both obvious and frustrating: Start over—get another DNA analysis, by a different and more careful laboratory. But where? Graciela herself was trained to work with challenging DNA samples, but she didn't even have a desk or a chair yet, let alone a sophisticated laboratory. Luckily for us, one of her Ph.D. colleagues, Jason Eshleman, was also a molecular anthropologist—a particularly good one, she said. After earning his doctorate, Jason had co-founded a company called Trace Genetics, which quickly established a reputation for extracting DNA from the most challenging samples possible: bones thousands of years old. "If anybody can get DNA for you," said Graciela, "it's Jason." I thanked her for the help and vowed to contact him.

Meanwhile, there was one final technique I wanted to try using the resources available to me at UT. It would require dismantling Joanna's beautiful reconstruction, and that seemed a shame. She'd spent weeks creating the sculpture, and what I wanted to do would destroy it in a matter of minutes. But there was nothing more that the reconstruction itself could tell us; besides, we had plenty of photos of it, taken from virtually every angle except the back of the head. Joanna began peeling off the clay with remarkable cheerfulness, and soon I found myself confronted once more by the bare skull. It looked smaller

dissolve to the skull. As the bone began to show through, I told him to stop. As I studied the superimposed images, I began to feel the same tingle of excitement I had felt when I first saw Joanna's facial reconstruction. Clearly visible through the flesh of Leoma Patterson's strong, rounded chin was the center of the skull's strong, rounded chin—the facial landmark named "pogonion." The cheekbones and the jaw lined up remarkably well, too. Near the top of the face, the skull's nasal bone ran directly beneath the midline of Leoma's nose, and the cranial landmark known as "nasion"—the indentation where the nasal bone meets the forehead—aligned precisely with the dip between Leoma's eyes.

Just weeks before, a DNA lab report had me convinced me that the skull was not Leoma Patterson's. Now I felt myself doing another U-turn. "It's her," I said. Half a dozen heads in the television studio nodded in agreement. I walked out of the studio feeling elated. The superimposition was powerfully persuasive.

And yet . . . And yet . . . It was persuasive, but it wasn't proof.

Our only hope for proof was to go back to the DNA. We'd have to try climbing the twisted ladder one more time. And by now I realized all too well how devilishly twisted that ladder could be.

For photos from Chapter 2, please visit: http://www.jeffersonbass.com/books/identitycrisis/ch2.html

Chapter Three

Twisted Ladder, Twisted Case

THE AIRPORT SECURITY screener peered closely at the two small objects in the envelope I had taken from my pocket. After a moment he called over another screener, and then another, and then a supervisor. Within minutes I was surrounded by armed Transportation Safety Administration guards who questioned me closely about the two human teeth I was carrying—my "carrion luggage," to make a terrible pun. Luckily, I was on my home turf—Knoxville's McGhee-Tyson Airport—and half the TSA guards there knew me. They weren't questioning me because they were suspicious; they were questioning me because they were curious. After I had explained why I was flying to Texas with two teeth from a dead woman in my pocket, they waved me on and wished me luck in San

Antonio, the next stop in my quest to determine, once and for all, whether 05-01 was Leoma Patterson after all. Like a runner in a relay race—a scientific race to find the truth—I was about to pass the baton to the next runner. In this case, the baton I was handing off consisted of the two teeth, and tucked inside those teeth, I hoped, was enough mitochondrial DNA to settle the matter.

Millions of people first learned of the forensic potential of DNA during the televised murder trial of O.J. Simpson. Blood from the scene where Nicole Simpson and Ronald Goldman were slaughtered—a sidewalk leading from Nicole's condominium—matched O.J. Simpson's blood type and DNA, a parade of prosecution witnesses testified. In addition, blood matching Nicole's DNA was found on a pair of socks in O.J.'s bedroom, and blood matching Ron Goldman's DNA was found in O.J.'s Bronco. The odds against anyone other than O.J., Nicole, and Goldman being the sources of those bits of blood-borne DNA were astronomical, experts explained, ranging as high as nearly one in ten billion.

The reason for such overwhelming mathematical probability, the jury learned during days of mind-numbing scientific testimony, is the immense size of the DNA molecule, coupled with the near-infinite ways in which its biochemical building blocks can be arranged. Its now-famous shape, called a double helix, resembles a ladder whose uprights have been twisted so they spiral or corkscrew around one another. The three billion "rungs" of this corkscrew ladder consist of pairs of chemical bases named adenine, guanine, cytosine, and thymine (abbreviated A, G, C, and T). For biochemical reasons I won't go into (because I don't begin to understand them!), every "rung" with an A on one side always—no exceptions—has a T on its other side; similarly, G is always paired with C. So if you were magically reduced to the size of an atom, and you began climbing this three-billion-rung ladder, the first five rungs you ascended might be constructed this way: AT, TA, CG, TA, GC. Think Legos—billions of Legos—spiraling miles into the sky.

Statisticians and evolutionary scientists have an old saying about monkeys and typewriters that's meant to illuminate how random variation, given enough time, produces specific meaning. One variation of that saying goes like this: Put a hundred million monkeys at a hundred million computer keyboards for a hundred million years, and one of them will write the complete works of William Shakespeare. Well, I don't actually believe that—I suspect it's far more likely that they'll bring the Internet and the entire computer-dependent world to its knees, and in far less than a hundred million years. But I can believe this, and can grasp it at a very rudimentary level: Stack up three billion pairs of ATs, TAs, CGs, and GCs, one atop another—or string three billion beads on a genetic necklace, or strike three billion two-note chords on a piano keyboard—and you're going to get one-of-a-kind combinations. Combinations named O.J. Simpson. Nicole Simpson. Ronald Goldman. Bill Bass. Leoma Patterson. All it takes to match a forensic DNA sample to a known individual—and to match it with rock-solid certainty—is an uncontaminated forensic sample, an uncontaminated sample from the known individual, and impeccable laboratory procedures. In the O.J. trial, the defense didn't challenge the scientific validity of DNA analysis, or of the matches, or of the statistical probabilities. Instead the defense argued—and the jury seemed to believe—that the incriminating samples were either planted by detectives or contaminated by careless lab technicians. So despite the staggering odds of guilt that the prosecution and its experts invoked—billions to one!—O.J. walked. The case hadn't been made beyond a reasonable doubt.

What about Leoma? Could a new DNA test indicate, beyond a reasonable doubt—in fact, with virtual certainty—whether the remains I had exhumed, not once but twice, were Leoma or some other murder victim? No one was on trial for murder in this case—Leoma's great-nephew had long ago confessed to killing her, and served six years for manslaughter (though in 2006 he recanted). In this case it wasn't a jury I was testifying before, but a family, one whose doubts I hoped to lay to rest forever.

In the O.J. Simpson case, the LAPD had been able to gather a fresh sample, within hours after the victims' death. In the Leoma Patterson case, unfortunately, the samples were more than twenty-five years old, and that was worrisome. DNA is immense, but it's also relatively fragile. Over time, it can break down or be destroyed by bacteria, acids, or other chemicals in the body or in the environment. Would enough DNA have survived with the tough, enameled capsules of the two molars nestled once more in my pocket at 34,000 feet as I streaked toward San Antonio? *Please*, I prayed silently.

SAN ANTONIO'S RIVERWALK, packed cheek-by-jowl with restaurants, bars, hotels, and tourists, is like a miniaturized, Disneyfied version of the canals of Venice or Amsterdam. Sprawling alongside the Riverwalk, the Henry B. Gonzales Convention Center encompasses 1.3 million square feet: roughly thirty acres, which is bigger than many family farms in East Tennessee. Toward the end of February 2007, those 1.3 million square feet were swarming with thousands of forensic scientists, converging on San Antonio for the annual convention of the American Academy of Forensic Sciences. The AAFS boasts more than six thousand members, in fields ranging from A (anthropology) to Z (zoarchaeology: "Doc, are those bones animal or human?"). The convention center's cavernous exhibit hall—bigger than the football field in Neyland Stadium—was jammed with hundreds of vendors' booths hawking a staggering array of goods and services: DNA analysis. Fingerprint scanners. Toxicology screenings. Disposable biohazard suits. Rape-test kits.

At Booth 117—sandwiched between two software companies—I met Jason Eshleman, a slight, soft-spoken, but swift-talking scientist. Jason is one of an emerging crop of twenty-first-century anthropologists inhabiting a niche I could not have imagined a half century ago: molecular anthropology.

Jason earned a Ph.D. at the University of California at Davis, under the tutelage of David Glenn Smith, who helped pioneer the application of modern DNA testing to ancient human skeletons. Smith enticed Jason to UC-Davis for his graduate work by posing an intriguing anthropological question: Were the dominant Native American tribes of the American Southwest—the Apaches and Comanches—the region's first human inhabitants, with a claim on the land dating back some ten thousand years? Or were the Apaches and Comanches interlopers, invaders who had swooped down from Canada only recently—"only" being about four thousand years ago? Smith told Jason that if he followed the trail of DNA evidence, he could answer that question. What he *didn't* tell Jason was that extracting and purifying ancient DNA was a nightmarishly difficult process. "I wouldn't wish it on my worst enemy," Jason laughed as he told me the story of his grueling research.

Trace Genetics, the DNA lab Jason co-founded after earning his Ph.D., had recently been acquired by DNA Print, a much larger firm, largely because of Jason's expertise in detecting and analyzing ancient DNA. One of his most notable successes had been finding DNA in the teeth of a ten-thousand-year-old skeleton from the Queen Charlotte Islands, located off the coast of British Columbia (an archipelago that bills itself as the "Galapagos of Canada"). Encouraged by this evidence of his skill, I handed over the envelope containing two molars I had pulled from the jaws of Maybe-Leoma.

Some scientists thrive on the conceptual; their minds can envision particles that the most powerful microscopes can't show us; processes that can't be directly observed, but only inferred, guessed at, by interpreting a stew of complex biochemical by-products. I am not one of these scientists. I need bones and teeth—things that I can see with my eyes and grasp with my hands. Jason Eshleman, on the other hand, can see with his mind's eye, grasping the complex interactions of the most complex molecules in the body, DNA.

One of the most famous, most challenging, and most controversial samples Jason had worked with was Kennewick Man, a male skeleton that a college student stumbled upon in the shallows of the Columbia River near Kennewick, Washington, in 1996. Initially believed to be the remains of a white homesteader, the bones were sent to a carbon dating lab just to be sure. The results—which put the age of the bones at 9,200 to 9,600 years old—ignited a fierce controversy. On one side were Native American tribes who wanted to rebury the remains as quickly and reverently as possible; on the other were scientists who wanted to study the bones for more clues about who first settled the Americas, where they came from, and how they got here. Several of my former Ph.D. students got caught up in the case, including George Gill, who teaches at the University of Wyoming; Doug Owsley, who heads the physical anthropology division at the Smithsonian Institution (and who got his bachelor's degree at Wyoming, under George Gill); and Richard Jantz, who directs the Forensic Center at UT (and who developed ForDisc).

Just months after the discovery of the bones—while Jason was still working in Smith's laboratory at UC-Davis—the lab received a small sample from the bones of Kennewick Man. Smith, Jason, and the other scientists were thrilled by the opportunity to do DNA analysis on the ancient remains, but before they had a chance to run the sample, the FBI came calling and took it. A consortium of Native American tribes had filed a legal motion to stop the analysis and recover the skeletal material, and a federal judge had granted an injunction and ordered the sample confiscated. Smith's lab complied, but not before putting the sample in a vial with a tamper-evident seal.

Nine years later, after scientists finally won the right to make a thorough study of Kennewick Man, Jason got the sample back, still sealed in the vial. To his surprise and disappointment, the only DNA he found in the sample proved to be quite a bit younger than nine thousand years, and it wasn't Native American. The DNA wasn't from Kennewick Man's, he finally discovered, but from another graduate student who had worked in the lab back in 1996. Jason didn't learn the genetic makeup of Kennewick Man, but he did learn a powerful lesson about how easily samples can be contaminated. It was a lesson underscored on other occasions by a mysterious, persistent contaminator. "There was a period when I kept seeing the same DNA sequence over and over in my samples," he recalls. "It wasn't from me, and it wasn't from anyone else who worked in the lab—we had everyone's sequence on file, so we knew it wasn't any of us." Eventually, the culprit emerged: Jason's girlfriend, who never set foot in the lab, but who shed enough stray skin cells to make her presence known, even though Jason always scrubbed up and suited up, much like a surgeon, before entering the lab.

As we talked amid the thicket of vendor booths in the cavernous exhibition hall in San Antonio, Jason explained how he would seek out and extract whatever DNA the teeth contained. He would start with just one of the teeth, keeping the other in reserve. His first step—and the reason he hadn't worried about my contaminating the teeth with my own DNA—would be to scrub the tooth with a solution of bleach to remove any surface dirt or other residue. Next, he would soak it in the bleach solution another five minutes. The bleach bath would destroy any DNA on the outside of the tooth, but five minutes wouldn't be enough time to allow the solution to penetrate the tooth's enamel and permeate the dentin, where the DNA would be ensconced. Next, he'd rinse the tooth with sterile water, then dry it under an ultraviolet lamp (another step designed to destroy any contamination on the surface). To gain access to the DNA within the molar, he would crack the tooth into smaller fragments and soak them in a solution that would dissolve the calcium, break down the proteins, and free the DNA from the cells. After the pieces were dissolved—a process that would take about a week, he said—he'd bind the DNA to silica beads, extract the digested proteins and other grunge, and finally wash the beads with an alcohol solution. "Just as alcohol washes away people's inhibitions," he explained, "it relaxes DNA's inhibitions, causing it to release from the silica beads."

That's when the real biochemical chemical wizardry would begin. Heating the solution almost to boiling causes the two legs of the DNA ladder to peel apart—like a zipper unzipping. When that happens, short (twenty-rung) pieces of complementary DNA called "primers" latch onto the long, unzipped halves of the mtDNA ladder, wherever their own sequences of A's, G's, C's, and T's mesh with the mtDNA's—a step called "annealing." The primers keep the ladder from zipping back together again as the solution is cooled slightly. Then, with the DNA unzipped, an enzyme called Taq (rhymes with "gack")—extracted from organisms that live in sulfur hot springs and hydrothermal ocean vents—moves in and builds a complementary new ladder-leg on each of the long unzipped legs . . . transforming one unzipped ladder into two ladders, and thereby creating two DNA strands out of one. That entire cycle, called a "polymerase chain reaction," or PCR, takes anywhere from twenty seconds to two minutes. At that point the entire solution is reheated to near-boiling again, the *two* ladders are unzipped, and the replication is repeated. It doesn't take many PCR cycles of unzipping and replicating— doubling the number of copies each time—to turn a few strands of DNA into many. Very, very many: In an hour or less, a single strand can be transformed into billions of copies.

By the time I finished talking to Jason, my head was spinning. I was in biochemical and genetic overload, but I felt sure the work and the teeth were in good hands. I wished Jason good luck in his search for DNA in the teeth.

My molar handoff wasn't the only casework I did while I was in San Antonio and surrounded by forensic experts. I took advantage of a book signing—Jon and I were signing copies of one of our Body Farm novels—to conduct an informal poll. We showed the 106 people in our book-signing line two images—Leoma Patterson's photo, and Joanna's facial reconstruction—and asked, "Are these the same woman, or not the same woman?" Of the 106 people polled, seventy-two said, "Yes, the same"; thirty-four said, "No, not the same." It wasn't a rigorous experiment by any means, but it sure was interesting to see people struggle to decide, and to listen in as they debated aloud with themselves or colleagues before voting.

I also did one other bit of research before leaving San Antonio. In the exhibit hall, I bumped into Murray Marks—one of my former Ph.D. students, whom I'd hired to join the forensic faculty at UT. Murray has used video superimposition in forensic cases, and he's also done research on computerized facial reconstruction, so I was particularly interested in his opinion on the case. He studied the facial superimposition video—not just the freeze-frame, but the footage itself, as the image slowly dissolved from the photo to the skull, then back to the photo again. When I put the question to him, he looked up from the computer screen, his expression halfway between amusement and exasperation. "What, are you retarded?" he said. "Of *course* it's her." I hoped Leoma Patterson's relatives would be as easy to convince. Convincing them would be easy, of course, if the DNA in the teeth confirmed that the bones were Leoma's after all.

I didn't realize what a big "if" that would turn out to be.

THREE WEEKS AFTER I handed the teeth to Jason, he phoned with discouraging news. "I'm not getting any DNA from the sample," he said. The reason wasn't clear. "It might be that I'm getting a lot of interference from humic material"—chemicals from the dirt, he explained, junking up the reaction—"or it might be that there just isn't any DNA left." It was possible, he went on, that acids from the soil, mold from the damp coffin, or bacteria had gradually invaded the tooth and destroyed the genetic material.

I was dumbfounded; how was it possible for DNA to survive for ten thousand years in teeth in the Pacific Northwest, but not for thirty years in East Tennessee?

"I don't know," Jason admitted. "There's not a lot of data from that part of the country." So we were providing new, discouraging data? I failed to find much comfort in that. "There's an extraction I can perform to remove the humic acids," Jason offered. "It might also remove some of the DNA—if there *is* DNA—but it might leave enough behind for me to get a sequence." At this point we seemed to have nothing to lose by trying.

A nail-biting week passed. "I'm seeing some DNA," Jason finally reported. Hallelujah! My excitement was short-lived, though: He had identified some pieces of mtDNA, but not enough yet to stitch together an entire sequence. "I'm going to process the second tooth now," he said, "and I'm hoping that will give me enough for a whole sequence." I hoped so, too.

But my hopes were in vain. The second DNA extraction failed to yield a complete mtDNA sequence. So did a third, a fourth, and a fifth. I had sent Jason more teeth; I sent him a four-inch section of femur. I sent him a cheek swab from Leoma's granddaughter Michelle, for comparison. But there was nothing in the skeletal material to compare with Michelle's DNA. Why not—where had the DNA gone? The answer came during a phone call to Dr. Cleland Blake, the medical examiner who had recovered and examined the remains back in 1979.

The remains still had bits of soft tissue on them, so Dr. Blake cleaned them. He did this by simmering the bones—"for a day or two," he said—in water containing detergent and bleach. Lots of bleach. Jason applied bleach, briefly, to destroy any DNA contaminating the samples' surfaces; Dr. Blake, on the other hand, had stewed the bones in it, and the combined assault of heat and chemicals had almost certainly nuked the DNA. Dr. Blake hadn't known any better—remember, the bones were found many years before forensic DNA testing was available—but the odds that Jason would succeed suddenly looked very slim.

Finally, in late May—three months after I'd optimistically handed over those first two teeth in San Antonio—Jason called. It was time to pull the plug, he said. He'd done seven extractions, all without finding a usable DNA sequence. It was a bitter blow. Abandoning the DNA quest meant abandoning our hope of making a positive identification. Ironically, after nearly two years, countless hours, and the best forensic techniques we could apply—ForDisc, DNA, a clay facial reconstruction, an experimental computerized reconstruction, and a video superimposition—we were coming full circle, ending up right back where we began: in uncertainty and ambiguity.

I scheduled a meeting with Leoma's relatives for June 1, 2007, to brief them on our efforts, our difficulties, and our confidence that—despite the GenQuest report, and the lack of sufficient DNA to refute it—the original identification had been correct after all. Personally, I felt sure that 05-01 was Leoma Patterson—the facial superimposition had convinced me—but I knew my belief wouldn't satisfy some of the family members, and I was dreading the meeting.

Then, at ten o'clock the night before the meeting, I received some astonishing news from David Ray, the original TBI investigator. David had long since traded his TBI badge for a sheriff's badge—he'd been elected sheriff of Claiborne County, up near the Kentucky border—but he'd heard about the twists and turns the Patterson case had taken. Intrigued by our difficulty finding DNA, David had rummaged around in the dusty files stored in his basement. There, he'd found his old TBI file on Leoma Patterson, and—wonder of wonders—in the file, he found the hair mat that had been recovered at the death scene back in 1979. The hair, along with bits of dried scalp, was sealed in a TBI evidence bag, the seals intact. My heart began to race. Unlike the bones, the hair and scalp had not been simmered in detergent and bleach. We were back in the game.

The following afternoon I briefed the family on all we'd done, and on how everything we'd done since the GenQuest test—the flawed, contaminated, water-muddling GenQuest test—had supported the original identification of the remains as Leoma's. The family appreciated all the efforts we'd made, but they were understandably disappointed that we couldn't offer certainty. When I showed them the hair and scalp, though—which David Ray had brought down from Claiborne County that morning—their hopes soared again.

As they gathered around and watched, I slit the TBI evidence seal and opened the bag containing the hair and scalp. Snipping off a hank of hair, I sealed it in a Ziploc plastic bag, along with a bit of dried scalp, and overnighted the bag to Trace Genetics. Jason planned to divide the samples; he would analyze one-half, he told me, while his senior technician—working independently, in a separate lab—analyzed the other.

On June 18, 2007, Jason reported that both he and his technician had found plenty of DNA in the hair and scalp. For simplicity, they were looking only at mitochondrial DNA—a simpler, more durable form of genetic material than nuclear DNA. "What we don't know yet is whether it matches the granddaughter's," he said.

Eight days later—on June 26, 2007—he knew: The DNA in the hair and scalp, and the DNA in Michelle's cheek swab, were identical. What's more, it was an unusual variety of mitochondrial DNA, one distinguished by two mutations. Statistically, the chances of a random match were extremely low, said Jason, just one in fifteen thousand. Turned around the other way, that meant the odds that Michelle was indeed the dead woman's granddaughter—that the dead grandmother was Leoma

Patterson and not some Jane Doe—were fifteen thousand
to one: for all practical purposes, one hundred percent.
After nearly two years and a wild forensic roller-coaster
ride, we had identified Leoma Patterson at last, conclu-
sively and positively. Maybe now Leoma—that is to say,
Leoma's *family*—could finally rest in peace.

For photos from Chapter 3, please visit: http://www.
jeffersonbass.com/books/identitycrisis/ch3.html

Epilogue

Knoxville, Tennessee
Spring 2015

TEN YEARS HAVE now passed since Michelle Atkins called to ask me to exhume a grave and take samples of the woman who might or might not be her grandmother. Maybe I'm getting old—truth is, at eighty-six, I've *been* old for quite a while—but it's hard to believe it's been that long since Michelle Atkins first phoned me. Sometimes it seems like just yesterday that I was standing in the grave for the second time . . . and maybe the day before yesterday that I corkscrewed over Redoak Mountain for the first time.

But no, it's been a decade, and time has not stood still, except in a few cobwebbed corners of my mind. Up in a wooded corner of the Cumberland Mountains, Leoma Patterson's grave—a raised, red scab the last time I saw it—is now smooth and grassy, the angry wound fully healed. She's been buried three times, and so far, knock wood, the third time seems to be the charm. My hope is that her family's wounds are healed, too, helped by the salves of time and a sense of closure.

Jimmy Ray Maggard, who confessed in 1985 to killing his great-aunt in a dispute over drugs, recanted his confession when he got wind of the problematic GenQuest report. By then he'd already served his six-year sentence for manslaughter, but he argued that the confession—which he claimed to have given under duress—was a key reason that he was sentenced to life imprisonment for the Georgia murder with which he was charged. Maggard filed a legal motion for postconviction relief—a reduction in his sentence—but the motion was denied. As of this writing, he remains incarcerated in Hays State Prison, some thirty miles south of Chattanooga, Tennessee. Despite receiving a life sentence, he is eligible for parole, but so far, he has not been granted it.

Officially, I was already retired from the Anthropology Department in 2005 when the case began, though I seem to have a problem *staying* retired. Since 2005, Richard Jantz has also retired, and Lee Jantz is getting close. Graciela Cabana, the molecular anthropologist who didn't even have an office chair when she sat down to help us find the flaws in the GenQuest analysis, now has three DNA labs in Stadium Hall: a modern DNA lab, an ancient DNA lab, and a forensic DNA lab.

Joanna Hughes, whose astonishing facial reconstruction put the investigation back on track after GenQuest derailed it, still takes on forensic cases from time to time. Since 2002 she's done a total of seventeen forensic reconstructions; of those, five have led to identifications. These days, though, Joanna's time is largely devoted to another project: raising the two young daughters she and her husband have.

Jon Jefferson—who didn't just write about the Leoma Patterson investigation, but helped advance it by asking key questions and prodding strategically—has written nine Body Farm novels in collaboration with me now. A resident of Knoxville when Leoma's case began, Jon pulled up stakes and moved to Baltimore, then back to Knoxville, and next to Tallahassee. (Jon jokes that he likes to stay one step ahead of the authorities.) Bitten by the forensic bug, Jon helped catalyze a major forensic investigation in Florida, leading to the exhumation of fifty-five graves on the grounds of a notorious reform school, the Dozier School for Boys, where—over the course of a century—stories of savage beatings, sexual abuse, and even killings had filtered out of the Florida panhandle town where the school was located. But that's another story, for another time. At the moment, Jon is still ensconced in Tallahassee, but he's starting to shop for a house in Athens, Georgia, where his wife—his third (sequentially, not simultaneously!)—is joining the faculty at the University of Georgia. As with Leoma's burials, so with Jon's marriages: The third time's the charm.

Since my first, life-changing forensic case some

fifty—no, sixty!—years ago, I've worked hundreds more. Some stand out more than others. The Leoma Patterson case is one of those, and always will be: fascinating, frustrating, and finally deeply rewarding. The case was a two-year roller-coaster ride, as twisted as DNA itself, as dizzying as that narrow corkscrew road over Graves Gap.

I didn't ride that roller coaster alone. A lot of smart, dedicated people rode it with me: Leoma's family. Investigators and prosecutors. Skilled scientists. A gifted artist.

Solving the mystery of Leoma Patterson took technology, but it also—especially—took teamwork. So does every forensic case. No matter how sophisticated the machines get, it's the *people* who are crucial, and always will be.

If you like a good forensic puzzle, you will love the Body Farm series by Jefferson Bass. Forensic anthropologist Dr. Bill Brockton is the fictionalized version of Dr. Bill Bass himself, and he uses all of his forensic expertise to get out of some pretty precarious situations. Don't miss Dr. Brockton's next adventure in . . .

The Breaking Point

Coming soon from William Morrow

Prologue

Friday, June 18, 2004
Knoxville, Tennessee

McCREADY STOPPED AND knelt beside a rut in the dirt road, raising a hand to halt the six men and two women fanned out behind him. The road—if a pair of faint tracks through grass, weeds, and leaves could indeed be called a road—meandered down a hillside of oaks and maples, their trunks girdled with vines. The mid-June morning was sweet with honeysuckle blossoms; the exuberant lushness of June had not yet given way to the duller green of July and the browning scorch of August, but underneath the perfume lurked something darker, something malodorous and malevolent hanging in the air.

McCready—Special Supervisory Agent Clint "Mac"

McCready—studied the rut, which was damp and also deeply imprinted with multiple layers of sharply defined tire tracks. He pulled two evidence flags from a back pocket and marked the ends of the tracks, then, with the camera slung around his neck, took a series of digital photographs. The photos were wide-angle views at first, followed by tighter and tighter shots. As he snapped the final, frame-filling close-ups, he said, to no one in particular, "It rained, what, couple days ago?"

"Night before last." The answer came from behind him, from Kimbo—Kirby Kimball, the youngest, newest, and therefore most eager member of SSA McCready's Evidence Response Team. "The front passed through about thirty-six hours ago. Rain stopped shortly after midnight."

McCready nodded, smiling slightly at the young agent's zeal, and lowered the camera, focusing now solely with his eyes. "These tracks look like they've been *machined*. What does that tell us?"

"New tires," said Kimball. "Deep tread blocks. Almost no wear. But there's a nick—a cut—here. At the outer edge."

"What else?"

"Big, off-road tires," Kimball added, squatting for a closer look. "SUV or four-by-four. Just one, looks like. One set of impressions heading in, another—on top— heading back out."

"Right." McCready glanced over his shoulder at the other agents. "Mighty quiet back there. I thought maybe the rest of you guys had gone for coffee." The agents exchanged sheepish glances. "Okay, what else can we tell from these tracks? Somebody besides Kimbo jump in. Anybody?"

"The vehicle passed through after the rain stopped." This from Boatman, an earnest, thirty-something agent who looked and listened a lot more than he talked.

"Right, far as it goes. But can you pin it down any tighter than that?"

Boatman stepped forward and bent down, his brow furrowing, his gaze shifting from the tracks to the surrounding vegetation—crabgrass and spindly poison ivy. "Quite a while after the rain stopped. Hours later, I'd say; maybe yesterday afternoon or even last night."

"Because?"

"The impressions wouldn't be so crisp—so perfect—if there'd been a puddle there when the vehicle went through," Boatman said. He surveyed the margins of the rut, then inspected the undersides of some of the blades of grass there. "Plus, if there'd been standing water, there'd be mud spatter on the vegetation. There's no spatter."

"Good." McCready focused on Kimball, who stood motionless yet somehow seemed cocked and ready to fire: his T-shirt stretched by the tension in his shoulders and biceps; the heels of his boots hovering a half inch off the ground, as if he were ready to spring into action. "Kimbo, you're an eager beaver this morning; you wanna cast these?" It wasn't actually a question.

"Yessir. On it." Kimball jogged back to the truck, a Ford Econoline chassis with a big cargo box grafted behind the cab; the vehicle might have passed for an ambulance on steroids if not for the prominent FBI logo on the side and the foot-high letters reading: EVIDENCE RE-SPONSE TEAM. Opening a hatch on the side of the vehicle, Kimball hauled out a large tackle box and lugged it to the tracks. He unlatched the lid and took out a gallon-sized Ziploc bag, half filled with powdered gypsum crystals—dental stone—and a graduated squeeze bottle. Squirting ten ounces of water into the bag, he resealed it and began kneading, creating a slurry the color and consistency of thin pancake batter: runny enough to flow into every block and groove of the tire tracks, thick enough not to seep into the soil itself.

McCready had already moved on, following the tracks in a hunched-over crouch: half bloodhound, half Quasimodo. "Looks like they parked here," he said, stopping to study the ground again. The soil was covered with leaves, and McCready frowned at the lack of castable shoe impressions. A trail of scuffed leaves led toward the trees at the edge of the clearing, but the undergrowth beyond the tree line appeared to be undisturbed; indeed, the scuff marks led only as far as a large, convex oval of mussed leaves situated just short of the trees. McCready began circling the oval, pausing occasionally to take photos.

"This matches the C.I.'s description of where it went down," he said. Heads nodded in agreement; earlier, McCready had passed out transcripts of his interview with the confidential informant. "Boatman, you and Kimbo . . ." He paused to glance over his shoulder at Kimball, who had already finished pouring the slurry of dental stone into the rut. "You and Kimbo set up the total station and start mapping. Rest of you, suit up and get ready to dig in."

The other six team members returned to the truck and wriggled into white biohazard suits and purple gloves. They came back laden with rakes, shovels, trowels, plastic bins, and a wood-framed screen of quarter-inch wire mesh.

As they laid their tools neatly beside the oval mound, Boatman latched the 3-D mapping unit onto a tripod. Kimball returned to the tire tracks again, this time holding a long, reflector-topped rod, its length marked in alternating twelve-inch bands of red and white. Boatman swiveled the instrument toward Kimball and sighted on the reflector. "Lights, camera, *action*," he deadpanned, and began pressing buttons to capture the position of the track. Checking the small display screen, he nodded. "Got it," he said, rotating the unit toward the oval mound, to which Kimball jogged with the reflector.

The mound, uncovered by careful raking, was red-brown clay, roughly four feet by six feet. The clay was broken and infused with pale, shredded roots, freshly shorn and torn from the soil—a raw, ragged wound in the earth's smooth, dark skin. McCready's gaze ranged over the lumpy surface, then zoomed in on something no one else had seen, tucked beneath a clod of clay. Kneeling just outside the margin of the oval mound, he leaned down, his nose practically in the dirt. "Cartridge case," he said. "That was careless of somebody." Then, without looking around: "Kimbo." By the time he'd finished saying the name, Kimball was already placing the end of the rod beside the piece of brass.

"Got it," Boatman called a moment later.

Still kneeling, McCready took a twig from the ground and used it to lift the shell from the clay. Angling it to catch the light, he peered closely at the marks in the base. "Remington. Nine millimeter." A paper evidence bag materialized beside his knee, held open by one of the agents; McCready dropped the case into it, and the agent sealed and labeled it, then set it in one of the plastic bins.

He sat back on his heels. "All right. We're burning daylight, so let's get to it. Boatman, you and Kimball keep mapping. The rest of you, dig in: Shovel till you see something, then switch to trowels. Screen everything—dirt, leaves, twigs, everything but the air. Hell, screen the air, too." He waved a hand in a sweeping gesture that encompassed not just the mound of clay but the surrounding area as well. "Might be more brass, buried or scattered around the periphery. Maybe cigarette butts, too, if we're lucky or the shooters are stupid. Maybe they left us some DNA."

"Maybe a signed confession, too," joked one of the agents. McCready did not laugh, so no one else did either.

"All right," he said. "Dig in. Easy does it, though. If our C.I.'s playing straight with us, we've got three bodies here—the two buyers and our undercover guy. Way the C.I. tells it, the traffickers never intended to sell; their plan all along was to kill the buyers, keep the coke, and move their own distributors into the dead guys' turf."

"Nice folks," muttered someone.

"Aren't they all?" someone else responded.

THEY BEGAN BY defining the margins of the grave with probes—thin, four-foot rods of stainless steel, each topped by a one-foot horizontal handle. Pressed into the soft earth of a fresh grave, the slender shafts sank easily; encountering hard, undisturbed soil, though, they balked and bowed, resisting. The probes weren't actually necessary; the perimeter of the grave was clearly visible, once the leaves and the slight mound of excess fill dirt had been removed. Still, the Bureau prided itself on thoroughness, and McCready was a Bureau man all the way. There would be no shortcuts today, for himself or his team.

Once the grave's outline was flagged and mapped and photographed, three of the agents—already sweating inside their biohazard suits—began digging. They started with shovels, working at the margins, digging down a foot all the way around before nibbling their way toward the carnage they expected to unearth at the center. After a grim twenty minutes, marked mainly by labored breathing and the rasping and ringing of shovel blades against soil and rocks, one of the agents—Starnes, a young woman whose blond hair spilled from the hood of her moon-suit like a saint's nimbus—paused and leaned in for a closer look. "Sir? I see fabric. Looks like maybe a shirtsleeve."

McCready knelt beside her. With the triangular tip of a thin trowel, he flicked away crumbs of clay. "Yeah. It's an arm. Lose the shovels. Switch to trowels. Let's pedestal the remains."

Two sweaty hours later, digging downward and inward from all sides, they'd uncovered a tangle of limbs, torsos, and heads. The pedestaled assemblage resembled a macabre sculpture—a postmortem wrestling match, or a pile of tacklers on a football field. It also reminded McCready, for some odd reason, of an ancient Roman statue he'd seen years before, in the Vatican Museums: a powerful sculpture of a muscular man and his two terrified sons caught in the crushing coils of sea serpents. Maybe the reason wasn't so odd after all, he realized: Like the chilling figures frozen in stone, these three men had died in the coils of something sinister, something that had slithered up behind them as surely and fatally as any mythological monster.

McCready photographed the entwined bodies from every angle, seemingly oblivious to the stench that grew steadily stronger as the day—and the corpses—got hotter. "All right," he said finally. "Give me three body bags over on this patch of grass. Let's lift them out one at a time. I'll want pictures after each one."

It took another half hour to lay out the corpses, faceup, on the open body bags. By then several of the techs were looking green around the gills, though no one had vomited. The last of the bodies to be lifted from the grave—the eyes gone to mush, the cheeks puffed out—was recognizable, just barely, as the man whose photograph McCready had passed around in the morning briefing. "This one's Haskell, our undercover guy," he said grimly.

"So the C.I. was telling us true," said Kimball. "The drug buy goes bad, turns into a shoot-out."

"Looks like it," said McCready. "But just to be sure, let's ask him." He turned, looking over one shoulder toward the trees on the far side of the clearing. "*Hey,*" he called out. "You—Brockton. Step out from behind that tree. And keep your hands where I can see them."

The team turned as a man emerged. He did not appear to be a seedy specimen from the sewers of the drug-trafficking world. The man looked more bookish than dangerous, and as he raised his hands, a broad smile creased his face.

Chapter One

"You—Brockton," I heard McCready calling. "Step out from behind that tree. And keep your hands where I can see them."

"I'm unarmed," I yelled, stepping from my observation post behind an oak tree. "But I've got a Ph.D., and I'm not afraid to use it. One wrong move and I'll lecture you to death!" The joke—*mostly* a joke—drew laughs from the weary FBI agents, as I'd hoped it would. "I'm Dr. Bill Brockton," I added as I approached. "Welcome to the Body Farm." I approached the rim of the empty grave, which was ringed with evidence flags and sweat-drenched FBI forensic techs. Peering into the hole, I saw that they had excavated all the way down to undisturbed soil, four feet down. The clay there was deeply grooved, as if it had been clawed by an immense monster. I, in fact, was that monster, and I'd left those marks the day before, when I dug the grave with a backhoe.

I'd missed most of today's excavation, having spent the morning entombed deep inside Neyland Stadium, the colossal cathedral to college football that the University of Tennessee had erected beside the emerald waters of the Tennessee River. Wedged beneath the stadium's grandstands, caught in a spider work of steel girders, was Stadium Hall: a dingy string of offices, classrooms, and laboratories, most of them assigned to the Anthropology Department, which I chaired. The rooms were strung along one side of a curving, quarter-mile corridor, one that underscored the *hall* in Stadium Hall. At midafternoon, when McCready had texted to say that the training exercise was nearly finished, I hopped into my truck, crossed the bridge, and slipped through a high wooden gate and down through the woods, stepping carefully to avoid treading on the bodies and bones scattered throughout the three-acre site: donated corpses whose postmortem careers were meticulously scrutinized, itemized, and immortalized, in photos, journal articles, scholarly dissertations, and law-enforcement anecdotes.

Officially, my macabre laboratory was named the Anthropology Research Facility, but a few years before, one of McCready's waggish FBI colleagues had dubbed it the "Body Farm," and the moniker—popularized by crime novelist Patricia Cornwell—had caught on so thoroughly that even I, the facility's creator, tended to call it by the catchy nickname. For several years now the FBI had been sending Evidence Response Team members to the Body Farm for training exercises like this one. With a ready supply of actual human corpses, plus plenty of privacy, the facility was the only place in the nation—possibly in the entire world—where forensic teams could hone their skills in such realistic scenarios.

The three corpses just unearthed by McCready's team had gradually attracted a cloud of blowflies, some of which strayed—either at random, or in an excess of eagerness—from the faces of the dead to the eyes and nostrils of the quick, causing the agents to squint and swat at the unwelcome intruders. Off to one side was a large mound of sifted dirt, plus piles of clay clods and rocks too big and too hard to pass through the quarter-inch wire mesh. On the ground beside the dirt lay the screen and—atop the mesh—three cartridge cases, two cigarette butts, and one wad of chewing gum, plus a gum wrapper.

I scrutinized the screen, then the bodies, then the hole in the ground, taking my time before turning to face the assembled agents. "That's it? That's all you got?" Their expressions, which had been confident and proud a moment before, turned nervous when I added, "So y'all just ran out of steam before you got to the fourth body?" Exchanging worried glances, they returned to the edge of the grave, their eyes scanning its floor and walls. I chuckled. "Kidding," I said, and a chorus of good-natured groans ensued. "Okay, so tell me what you've learned from the scene."

I pointed at Kimball, the eager young agent who'd cast the tire tracks. "Agent Kimball," I said. "You like to make a good . . . *impression*." More groans, as the dreadful pun sank in. "What else does that rut tell us, besides the fact that the puddle had dried up by the time the tracks were made?" McCready had texted me a few notes on the team's findings, starting with their observations about the tire impressions. Kimball frowned, so I gave him a hint. "How many sets of tracks did you cast?"

"Just the one," he said. "That's all . . ." He hesitated, his eyes darting back and forth, then the light dawned. "Ah—they all rode in together."

"Bingo," I said. "But they didn't all ride out together. And what about the grave? What does the evidence there tell us?"

"The cartridge cases are from two different weapons," said one of the dirt sifters. "They're all nine-millimeter Remington, but there's two different firing-pin impressions. One's round, the other's rectangular." I nodded approvingly; when I'd asked a friend on the campus police force for spent shells, I specifically requested shells from two different handguns, so I was pleased that the difference had been noticed. "Also," he went on, "the cigarette butts are two different brands. So we might get two different DNA profiles from those."

"Good," I said. "Maybe there's DNA in the gum, too—and maybe the gum chewer's not one of the smokers. So there could be *three* DNA profiles, right?" Heads nodded. "Okay, let's talk taphonomy—the arrangement of the items you excavated. What did you learn as you unearthed the bodies?"

"All three were killed with a single shot to the back of the head," said a guy whose nerdy, Coke-bottle glasses were offset by immense, chiseled muscles, gleaming with sweat and smears of clay. "Execution style." I nodded, slightly self-conscious about this part. The shots to the head were the least realistic part of the exercise, because the shots—unlike the corpses themselves—were fakes. It had struck me as unnecessary and disrespectful to fire bullets into donated bodies, so I'd settled instead for daubing a small circle of red dye onto the back of each head, and a larger circle on each forehead, to simulate entry and exit wounds.

"What else?" A long silence ensued. "Did you find blood in the grave?" Heads shook slowly. "Did you find blood *anywhere* besides on the wounds themselves?" More head shaking; several of the agents now cast nervous sidelong glances at one another. "So what does that suggest to you?"

The blond woman raised a hand. "It suggests they were killed somewhere else," she said. "And then brought here."

I gave her a thumbs-up. "Which explains why there was only one vehicle. Tell me—how often do drug traffickers and drug buyers carpool to the place where the deal's going down?" A few of the agents laughed, but Kimball, the tread caster, winced, as he should have: Kimball, of all people, should have given more thought to the absence of a second vehicle. "Also," I went on, "how likely is it that only three bullets would be fired during a drug-deal shoot-out? All of them to the back of the victims' heads?" I could see them rethinking the scenario. "Anything else?" The agents looked from the grave to the bodies and back to the grave, then at me once more.

My questions made it clear that they were still missing something—still failing to connect important dots—but apparently they needed a hint. "Look closely at the three faces," I said. "See any differences?"

"Ah," said the nimbus-haired blonde. "The two 'buyers' look a lot better than our guy. A lot . . . *fresher.*"

"Bingo," I said. "They show no signs of decomposition, and no insect activity. Look at your 'undercover agent.' He's a mess—he's starting to bloat, and he's got maggots in his mouth and nostrils. Anybody look in there?" Several of the agents grimaced; most shook their heads sheepishly. "So if you compare the condition of the bodies, what does the difference in decay tell you?"

"He was killed before the other two," said Boatman, the agent who'd noticed the absence of mud spatter beside the tire tracks.

"Exactly," I said, pulling on a pair of purple nitrile gloves. "Also, your undercover guy was probably outdoors, or maybe stashed outdoors for a while—someplace where the blowflies could get to him." I pointed a purple finger at the puffed-up face again. "Blowflies like to lay their eggs in the moist orifices of the body," I went on. "The mouth, the nose, the eyes, the ears, even the genitals, if those are accessible. But especially, *especially,* any bloody wound." I stooped beside the dead "agent" and lifted his head. I had gone to the trouble of mixing a bit of actual blood—pig blood—with the red food coloring on his head, and I'd brought him out to the Body Farm two days before I brought the other bodies. During that time, his "gunshot wound" had attracted legions of flies, and by the time I'd placed the bodies in the ground, maggots had begun colonizing his hair, forehead, and orifices. "Next time, check for maggots. And collect the biggest ones." I bent down and plucked a quarter-inch specimen from an eye socket, holding it in my palm for them to inspect. "A forensic entomologist could tell you that this maggot hatched three or four days ago. Which—if I remember right—is just about the time your undercover agent dropped off the radar screen. Is that correct, Agent McCready?"

"That's correct, Dr. Brockton."

I flicked the maggot into the woods. It was time to reveal the final plot twist in the scenario. When I first phoned to suggest the idea, McCready had sounded dubious. As we talked, though, he warmed to the idea, and by the end of the call, he'd embraced the scenario enthusiastically. "A good lesson in investigative skepticism," he'd called it.

"So," I said to the team of trainees, "knowing that these other two guys were killed a couple days after your agent— and knowing that all of them were brought out here and buried together . . ." I paused, giving them time to think and rethink before offering the final hint. " . . . what does that tell you about your confidential informant?"

"It tells us he's a lying sack of shit," Kimball blurted. His face was flushed and his tone was angry, as if the corpse really *was* a murdered FBI agent, rather than a married insurance agent who'd had a heart attack during a tryst with his mistress. "It tells us the C.I.'s whole story is bullshit," Kimball fumed, smacking a fist into an open palm. "Hell, maybe he even set *up* our guy—ratted him out to the traffickers."

I nodded. "Maybe so. So be careful who you trust. Bad guys lie through their teeth. But bugs?" I pointed to the bloated face and the telltale maggots. "You can always believe them. Whatever they tell you, it's the truth."

Chapter Two

THE FAMILIAR ARC of a rib cage filled my field of vision as I leaned down and peered through the smoke. On the rack of my charcoal grill, two slabs of baby-back ribs sizzled, the meat crusting a lovely reddish brown. Ribs were a rare treat these days—Kathleen, invoking her Ph.D. in nutrition, had drastically cut our meat consumption when my cholesterol hit 220—but she was willing to bend the dietary rules on special occasions. And surely this, our thirtieth wedding anniversary, counted as a special occasion.

As soon as the FBI training at the Body Farm ended, I'd headed for home, stopping by the Fresh Market, an upscale grocery, to procure the makings of a feast, southern style: ribs, potato salad, baked beans, and coleslaw.

As I fitted the lid back onto the smoker, I heard a car pull into the driveway, followed by the opening and slamming of four doors and the clamor of four voices.

A moment later the backyard gate opened and Jeff, my son, came in. Leaning into the column of smoke roiling upward, he drew a deep, happy breath. "Smells great. Almost done?"

"Hope so. The guest of honor should be home any minute. She's been dropping hints all week about celebrating at the Orangery." The Orangery was Knoxville's fanciest restaurant. "Way I see it, only way to dodge that bullet is to have dinner on the table when she gets here."

"You know," Jeff said, "it wouldn't kill you to take Mom someplace with cloth napkins and real silverware once every thirty years."

I raised my eyebrows in mock surprise. "You got something against the plastic spork? Anyhow, I thought it'd be nicer to celebrate here."

The wooden gate swung open again—burst open, whapping against the fence—and Tyler came tearing into the backyard, with all the exuberance of a five-year-old who'd just been liberated from a car seat. "Grandpa *Bill*, Grandpa *Bill*, I could eat a *horse*," he announced, wrapping himself around my left leg.

A few steps behind came his younger brother, Walker, age three, grabbing my right leg and crowing, "I can eat a elephant!"

Jeff's wife, Jenny—a pretty, willowy blonde, who carried herself with the easy grace of an athlete—came up the steps after them, closing the gate. "Stay away from the grill, boys," she called. "It's hot. Very, very hot." She leaned over the boys to give me a peck on the cheek. "I don't know about the ribs, but you smell thoroughly smoked," she said. "Are you *sure* you want us horning in on your anniversary dinner?"

"Absolutely. What better way to celebrate thirty years of marriage?"

"Hmm," Jeff grunted. "Hey, how 'bout you and Mom celebrate with the boys while Jenny and I eat at the Orangery?"

"Listen to Casanova," scoffed Jenny. "For *our* anniversary, he took me to the UT-Vanderbilt game. *Super-romantic.*" She shook her head good-naturedly. Then, with characteristic helpfulness, she asked, "What needs doing?"

"If you could set the table, that'd be great. Oh, and maybe put the slaw and potato salad and beans in something better looking than those plastic tubs?"

She nodded. "Hey, kiddos, who wants to be Mommy's helper?"

"*I* do, *I* do," they both shouted, abandoning me to follow her through the sliding glass door and into the kitchen.

"What on earth did you do to deserve her?" I asked Jeff as the door slid shut.

"I think she likes me for the foil effect," he said. "I make her look so good by comparison. Same reason Mom keeps you around."

I heard the quick toot of a car horn in the driveway then, followed by the clatter of the garage door opening. Kathleen was home.

Soon after, delighted squeals—"Grandmommy! Grandmommy!"—announced her arrival in the kitchen.

The slider rasped open and she emerged, her leather briefcase still slung over her shoulder. "Bill Brockton, you *sneak*. You didn't tell me you were cooking."

"I wanted to surprise you."

"I wanted to surprise you, too," she said. "I made us a seven o'clock reservation at the Orangery."

"Oh, darn—I wish I'd known," I said. She shot me a dubious look, which I countered with an innocent smile. "That would've been nice, honey. But I guess you'd better call and cancel."

"I'll call," she said, "but I won't cancel; I'll reschedule, for Saturday night. You don't get off the hook *that* easily. If I can survive thirty years of Cracker Barrel vittles, one fancy French dinner won't kill you."

She turned and headed inside. The instant the sliding-glass door closed, Jeff and I looked at each other and burst into laughter.

Dinner was loud, rowdy, and wonderful, with three terrible puns (all of them mine), two brotherly squabbles, and one spilled drink (also mine). The ribs were a hit—smoky, succulent, and tender.

Sitting at the head of the kitchen table, I surveyed my assembled family, then, with my sauce-smeared knife, I tapped the side of my iced tea glass. "A toast," I said. The three adults looked at me expectantly; the two boys gaped as if I were addled.

"Toast?" said Walker. "Toast is breakfast, silly."

"A toast," Jenny explained, "is also a kind of blessing. Or a thank-you. Or a wish."

Walker's face furrowed, then broke into a smile. "I

toast we get a dog!" His toast drew laughs from Kathleen and me, and nervous, noncommittal smiles from his parents.

"A toast," I repeated. "To my lovely wife. To thirty wonderful years together."

We clinked glasses all around. Kathleen looked into my eyes and smiled, but then, to my surprise, she teared up. "To this lovely moment," she said, her voice quavering, "and this lovely family. The family that almost wasn't."

Now I felt my own eyes brimming. We almost never spoke of it, but none of us—Kathleen, Jeff, Jenny, or I—would ever forget the near miss to which she was alluding. The grown-ups clinked glasses again—somberly this time—and Kathleen reached out to me with her right hand. Instead of clasping hands, though, she bent her pinky finger, hooked it around mine, and squeezed. It was our secret handshake, of sorts: our reminder of what a sweet life we had, and how close—how terribly close—we'd come to losing it, right in this very room, right at this very table, a dozen years before. I lifted her hand to my face and uncrooked her finger, tracing the scar around the base and then giving it a kiss. By now the scar was a faint, thin line—barely visible and mostly forgotten, except when something triggered memories of that nightmarish night, and that evil man: Satterfield, sadistic killer of women. Satterfield, emerging from our basement, gun in hand, to bind us—Kathleen, Jeff, me, and even Jenny, Jeff's girlfriend at the time—to the kitchen chairs. Satterfield, putting Kathleen's finger into the fishlike jaws of a pair of gardening shears and then closing the jaws in a swift, bloody bite.

Odd, how memories can open underfoot in the blink of an eye, taking you down a rabbit hole of the mind to some subterranean, subconscious universe where different rules of time and space and logic hold sway. Part of me remained sitting at the table, my fingers smeared with barbecue sauce, but part of me had gone down that bloody rabbit hole.

Kathleen's finger, which had sent me spinning there, now beckoned me back. She stroked my damp cheek and smiled again. "Will you marry me, Bill Brockton?" she asked.

"Yes, please," I answered. "Again and again. Every day." Half rising from my chair, I leaned over and kissed her—a grown-up kiss, on the mouth, taking my time.

"*Gross,*" said Tyler.

"Gross gross *gross,*" agreed Walker.

IT WAS TEN-THIRTY by the time Jeff's family was gone, the kitchen was clean, and Kathleen and I were showered and in bed. I rolled toward her on the mattress and cupped her face in one hand. "Not as romantic as the fancy French dinner you wanted," I said, "but tasty."

"Says the man who thinks turkey jerky is a delicacy," she said. "But yes, delicious. And it's always so sweet to see Jeff and Jenny with the boys. They're such good parents, Bill."

"They should be. You're a great role model."

"You, too," she said, and then, from nowhere: "You still sad we couldn't have more?"

"No," I said, though that wasn't entirely true; deep down, I would always wish I'd had a daughter as well as a son. "I'm the luckiest man alive. I couldn't be happier." I felt the stirrings of desire, and I slid my hand down to her hip. "Well, maybe I could be a *tiny* bit happier."

She smiled, but she also shook her head. Taking my hand from her hip, she brought it to her lips and gave it a consolation-prize kiss. "I need a rain check, honey. Bad time of the month."

"Still?"

She nodded glumly.

"That doesn't bother me," I assured her. "You know I'm not squeamish."

"I do know, and I appreciate that," she said. "But I'm just not up to it. I'm sorry, sweetie; I'll be off the sick list soon, and I *will* make it up to you. I promise."

She crooked her little finger at me again, to make sure I knew she meant it.

"I'm sorry it's giving you trouble," I told her, my disappointment giving way to sympathy. "Seems like that's gotten worse again. You need to go back to the doctor?" She'd had outpatient surgery a year or so ago, to remove a uterine fibroid—a knot of benign tissue—and her cramps and bleeding had lessened afterward. For a while.

"I think it's just menopause, letting me know it's headed my way," she said. "Now turn out the light and spoon me." She rolled over and snuggled against me.

Switching off the light, I wrapped an arm tightly across her chest. Her breathing slowed and deepened, her body twitching as she sank into sleep. As my own breathing found the same cadence as hers, I made a silent wish for her—one last anniversary toast, Walker style: *I toast you sleep well and feel better tomorrow.*

About the Author

JEFFERSON BASS is the writing team of Dr. Bill Bass and Jon Jefferson. Dr. Bass, a world-renowned forensic anthropologist, founded the University of Tennessee's Anthropology Research Facility. Jon Jefferson is a veteran journalist, writer, and documentary filmmaker.

www.JeffersonBass.com
www.witnessimpulse.com

Discover great authors, exclusive offers, and more at hc.com.

Chapter One

Reopening a Coffin,

Reopening a Case

FORTY MILES NORTHWEST of Knoxville—deep in the heart of hardscrabble coal country—Redoak Mountain nestles amid the peaks and valleys of the Cumberlands, an Appalachian range angling through Tennessee and up along the Kentucky-Virginia border. As mountains go, Redoak isn't particularly noteworthy. It tops out at 3,200 feet, less than half the height of the loftiest peaks in the nearby Great Smoky Mountains. Yet somehow—through some remarkable, unfortunate confluence of geology, bulldozing, and paving—the narrow switchback road I found myself navigating one muggy summer morning seemed to hang about 3,199 of those 3,200 feet above the steep, wild valleys below. My view of the valleys was clear and unobstructed. It was also more than a little unnerving, because the invention known as the guardrail—a common sight along many mountain roads in Tennessee—did not seen to have made its way yet into these perilous parts.

Even the place names out here, scarcely a stone's throw from my ivory tower at the University of Tennessee's Anthropology Department, harked back to an earlier, wilder world: Bearwallow Branch. Backbone Ridge. Graves Gap. One-room churches and tiny cemeteries seemed to outnumber the houses, unless you counted the handful of mobile homes, which appeared to have been airlifted into their notches in the steep hillsides. The route we were following over Redoak Mountain ("we" being writer Jon Jefferson, graduate student Kate Spradley, and I) was far off the beaten track. If I hadn't already noticed that fact on the high, hairpin curves, I'd have surely realized it when the guy we were following—a local who supposedly knew where we were going—got lost, taking us ten miles in the wrong direction before reversing course and turning up a side road.

The threadbare fabric of civilization clinging to the slopes and hollows of Redoak Mountain might logically raise the question, "Why?" Why bother carving roads and snaking trailers into these remote backwoods areas? The answer came from the heavy trucks groaning down the grades. The Cumberlands might be poorly settled, but they're richly mineraled, and the trucks—laden with strip-mined coal for the power plants of the Tennessee